Instructor's Solutions Manual

to accompany

Elementary Linear Algebra

FOURTH EDITION

Stewart Venit • Wayne Bishop
California State University at Los Angeles

PWS PUBLISHING COMPANY

I(T)P
An International Thomson Publishing Company
BOSTON • ALBANY • BONN • CINCINNATI
DETROIT • LONDON • MADRID • MELBOURNE
MEXICO CITY • NEW YORK • PARIS • SAN FRANCISCO
SINGAPORE • TOKYO • TORONTO • WASHINGTON

PWS PUBLISHING COMPANY
20 Park Plaza, Boston, MA 02116-4324

I(T)P™
International Thomson Publishing
The trademark ITP is used under license.

For more information, contact:
PWS Publishing Co.
20 Park Plaza
Boston, MA 02116

International Thomson Publishing Europe
Berkshire House I68-I73
High Holborn
London WC1V 7AA
England

Thomas Nelson Australia
102 Dodds Street
South Melbourne, 3205
Victoria, Australia

Nelson Canada
1120 Birchmont Road
Scarborough, Ontario
Canada M1K 5G4

International Thomson Editores
Campos Eliseos 385, Piso 7
Col. Polanco
11560 Mexico D.F., Mexico

International Thomson Publishing GmbH
Konigswinterer Strasse 418
53227 Bonn, Germany

International Thomson Publishing Asia
221 Henderson Road
#05-10 Henderson Building
Singapore 0315

International Thomson Publishing Japan
Hirakawacho Kyowa Building, 31
2-2-1 Hirakawacho
Chiyoda-ku, Tokyo 102
Japan

Printed and bound in the United States of America by Financial Publishing.
95 96 97 98 99—10 9 8 7 6 5 4 3 2 1

ISBN: 0-534-95191-0

This book is printed on recycled, acid-free paper.

Contents

Preface

This Instructor's Manual is intended for use with the text *Elementary Linear Algebra* (Fourth Edition) by Stewart Venit and Wayne Bishop. Its nine chapters are keyed to the corresponding ones in the text. Each chapter in the manual contains the following sections:

1. An overview of the corresponding chapter in the text. This overview provides a detailed list of the chapter's contents and should be useful in planning your lectures.

2. Remarks concerning the subject matter of that chapter in the text. These comments include suggestions for presenting the material, more detailed descriptions of some of the items listed in the overview, and a discussion of how the various topics tie in with the rest of the text (including which ones can be omitted if you are pressed for time). Together with the overview section, these remarks should come in handy in determining what material to include in your course.

3. Complete solutions to all the Self-Test problems and the odd-numbered Review Exercises. (The solutions to all Self-Test problems and *every other* odd-numbered Review Exercise also appear in the Student's Solution's Manual.)

4. Answers to the text's even-numbered exercises. (The answers to the odd-numbered exercises are given in the text itself.)

CHAPTER 1

Geometry of R^2 *and* R^3

Chapter Overview

Thc topics covered in this chapter are:

1. Vectors in $\mathbf{R}^2$ and $\mathbf{R}^3$ as ordered pairs or triples (represented by boldface, lowercase letters such as $\mathbf{v}$). [Section 1.1]
2. Equality of vectors, zero vector, and length of a vector in $\mathbf{R}^2$ and $\mathbf{R}^3$. [1.1]
3. Scalar multiples of vectors in $\mathbf{R}^2$ and $\mathbf{R}^3$; geometric interpretation. [1.1]
4. Sum and difference of vectors in $\mathbf{R}^2$ and $\mathbf{R}^3$; geometric interpretation. [1.1]
5. Properties of vector addition and scalar multiplication of vectors in $\mathbf{R}^2$ and $\mathbf{R}^3$. [1.1]
6. Directed line segments (represented as $\overrightarrow{\mathrm{PQ}}$); vector equivalent to a given directed line segment. [1.1]
7. Alternate ($\mathbf{i}$, $\mathbf{j}$, $\mathbf{k}$) notation for vectors in $\mathbf{R}^2$ and $\mathbf{R}^3$; for example, $2\mathbf{i} + 3\mathbf{j}$. [1.1]
8. Dot product of two vectors in $\mathbf{R}^2$ and $\mathbf{R}^3$; properties of dot product. [1.2]
9. Cross product of vectors in $\mathbf{R}^3$ (2×2 determinants are introduced to help with definition); properties of cross product. [1.2]
10. Point-normal and standard forms for a plane in $\mathbf{R}^3$. [1.3]
11. Point-parallel, two-point, and parametric forms of a line in $\mathbf{R}^2$ and $\mathbf{R}^3$. [1.3]

Remarks

Chapter 1 supplies a preview of, and a geometric feel for, Euclidean m-space ($\mathbf{R}^m$). The material presented here is usually covered in the calculus sequence. If calculus is a prerequisite for your linear algebra course, this chapter could be covered very quickly or skipped entirely. All the basic information concerning vectors in $\mathbf{R}^m$ that will be needed to read this text is given in Section 2.1 (which also acts a review of most of Chapter 1).

Most students seem to find the material in this chapter fairly straightforward. (In fact, the same can be said for much of the material in the first four chapters of the text.) For the most part, learning the material in Chapter 1 involves memorizing a relatively large number of definitions and then applying them to "mechanical" problems. Section 1.3, *Lines and Planes*, is perhaps the most difficult section in this chapter. Some students have difficulty visualizing the situation here, and needless to say, drawing lots of pictures when you cover this material should help considerably.

Solutions to Self-Test

1a.
$$\begin{aligned} 2\mathbf{y}\cdot\mathbf{z} &= 2(3, 2, 1)\cdot(-2, 3, -1) \\ &= (6, 4, 2)\cdot(-2, 3, -1) \\ &= (6)(-2) + (4)(3) + (2)(-1) \\ &= -2 \end{aligned}$$

1b. Let $\mathbf{y}\times\mathbf{z}$ be the vector (a, b, c). Then:

$$a = \det\begin{bmatrix} 2 & 1 \\ 3 & -1 \end{bmatrix} = (2)(-1) - (1)(3) = -5$$

$$b = \det\begin{bmatrix} 1 & 3 \\ -1 & -2 \end{bmatrix} = (1)(-2) - (3)(-1) = 1$$

$$c = \det\begin{bmatrix} 3 & 2 \\ -2 & 3 \end{bmatrix} = (3)(3) - (2)(-2) = 13$$

Thus, $\mathbf{y}\times\mathbf{z} = (-5, 1, 13)$.

1c.
$$\begin{aligned} \|3\mathbf{z}\| &= \|3(-2, 3, -1)\| = \|(-6, 9, -3)\| \\ &= \sqrt{(-6)^2 + 9^2 + (-3)^2} = \sqrt{126} = 3\sqrt{14} \end{aligned}$$

1d. $d(3\mathbf{y}, 6\mathbf{z}) = \|6\mathbf{z} - 3\mathbf{y}\| = \|(-12, 18, -6) - (9, 6, 3)\| = \|(-21, 12, -9)\|$

$$= \sqrt{(-21)^2 + 12^2 + (-9)^2} = \sqrt{666} = 3\sqrt{74}$$

2. The two-point form is given by:

$$\mathbf{x}(t) = (1 - t)\mathbf{p} + t\mathbf{q} = (1 - t)(1, -1, 1) + t(4, 3, 2)$$

The point-parallel form is obtained by multiplying out and combining like terms:

$$\mathbf{x}(t) = (1, -1, 1) - t(1, -1, 1) + t(4, 3, 2) = (1, -1, 1) + t(3, 4, 1)$$

3. The parametric equations of the line are:

$$x = t - 1, \quad y = 2t, \quad z = 4t + 1$$

Substituting these values of x, y, and z into the equation of the plane, $2x - y + z = 5$, yields:

$$2(t - 1) - 2t + 4t + 1 = 5.$$

We now solve this equation we obtain $t = 3/2$, and substitute this value of t into the parametric equations to find the point of intersection: $x = 1/2, y = 3, z = 7$, or $(1/2, 3, 7)$.

4. The normal to the given plane is $(2, -3, 1)$. Now, all vectors perpendicular to the given plane are scalar multiples of $(2, -3, 1)$, namely $\mathbf{x}(t) = t(2, -3, 1)$.

5. This line is parallel to the vector $(2, -3, 1)$, the normal to the plane. Thus, its equation is:

$$\mathbf{x}(t) = (2, 1, -3) + t(2, -3, 1)$$

6. A normal to the given plane is given by the vector $(1, -1, 2)$. A point on the plane can be found by setting $y = z = 0$, and solving its equation for x. This yields the point $(6, 0, 0)$. Thus, the point-normal form for the given plane is:

$$(1, -1, 2) \cdot (\mathbf{x} - (6, 0, 0)) = 0$$

7. The given lines are parallel to the vectors $(2, 1, 0)$ and $(1, -1, 1)$, respectively. Since these vectors are not scalar multiples of one another, they do not determine the same direction, and hence the given lines are *not* parallel.

8. Let $\mathbf{p} = (1, 0, 1)$, $\mathbf{q} = (1, 1, 2)$, and $\mathbf{r} = (3, 2, 0)$. We first find a vector that is normal to the plane determined by $\mathbf{q} - \mathbf{p}$ and $\mathbf{r} - \mathbf{p}$ by finding the cross product:

$$(\mathbf{q} - \mathbf{p}) \times (\mathbf{r} - \mathbf{p}) = (0, 1, 1) \times (2, 2, -1) = (-3, 2, -2)$$

Thus, the desired plane has equation $-3x + 2y - 2z = d$. To find d, we substitute the three coordinates of any of the given points into this equation. For example, using the first point:

$$d = -3(1) + 2(0) - 2(1) = -5$$

Thus, the equation of the desired plane is $-3x + 2y - 2z = -5$.

Solutions to Odd-Numbered Review Exercises

1.
$$\begin{aligned} \mathbf{u} - 3\mathbf{v} &= (1, 2, 0) - 3(-1, 0, 3) \\ &= (1, 2, 0) - (-3, 0, 9) \\ &= (4, 2, -9) \end{aligned}$$

3.
$$\begin{aligned} d(2\mathbf{u}, \mathbf{v}) &= \|\mathbf{v} - 2\mathbf{u}\| = \|(-1, 0, 3) - 2(1, 2, 0)\| \\ &= \|(-1, 0, 3) - (2, 4, 0)\| = \|(-3, -4, 3)\| \\ &= \sqrt{(-3)^2 + (-4)^2 + 3^2} = \sqrt{34} \end{aligned}$$

5.
$$\begin{aligned} \|-2\mathbf{u}\| &= \|-2(1, 2, 0)\| = \|(-2, -4, 0)\| \\ &= \sqrt{(-2)^2 + (-4)^2 + 0^2} = \sqrt{20} = 2\sqrt{5} \end{aligned}$$

7. Let $\mathbf{u} \times \mathbf{v}$ be the vector (a, b, c). Then:

$$a = \det\begin{bmatrix} 2 & 0 \\ 0 & 3 \end{bmatrix} = (2)(3) - (0)(0) = 6$$

$$b = -\det\begin{bmatrix} 1 & 0 \\ -1 & 3 \end{bmatrix} = -[(1)(3) - (0)(-1)] = -3$$

$$c = \det\begin{bmatrix} 1 & 2 \\ -1 & 0 \end{bmatrix} = (1)(0) - (2)(-1) = 2$$

Thus, $\mathbf{u} \times \mathbf{v} = (6, -3, 2)$.

9. $\mathbf{u} \cdot \mathbf{v} = (1, 2, 0) \cdot (-1, 0, 3) = (1)(-1) + (2)(0) + (0)(3) = -1$

11. The vector opposite to that of **v** is $-\mathbf{v} = (1, 0, -3)$. Hence, a unit vector in this direction is

$$\frac{-\mathbf{v}}{\|\mathbf{v}\|} = (1/\sqrt{10})(1, 0, -3) = (1/\sqrt{10}, 0, -3/\sqrt{10})$$

13. Since the z-axis is perpendicular to the plane, we can take (0, 0, 1) as a normal vector. Therefore, an equation of the plane (in point-normal form) is:

$$(0, 0, 1) \cdot (\mathbf{x} - (1, 2, 3)) = 0$$

or, in standard form, $z - 3 = 0$.

15. The coefficients of the variables give us a vector, (3, 1, -2), normal to the plane. To obtain a point on the plane, we set x and z equal to zero in its equation and solve for $y = 1$. Thus, a point on the plane is (0, 1, 0) and its point-normal equation is:

$$(3, 1, -2) \cdot (\mathbf{x} - (0, 1, 0)) = 0$$

17. The vector **v** not only gives the direction of the desired line but also gives a point (1, 0, 2) on it. Hence, the equation is:

$$\mathbf{x}(t) = (1, 0, 2) + t(1, 0, 2) = (1 + t)(1, 0, 2)$$

Setting $s = 1 + t$, we see that we can also write this equation more simply as $\mathbf{x}(s) = s(1, 0, 2)$. (This result can also be obtained by noticing that the origin,**0**, must also be a point on the line, and using it instead of (1, 0, 2) in the point-parallel form of the line.)

19. Recall that in the standard from equation for a plane,

$$ax + by + cz = d,$$

a, b, and c are the components of a vector normal to the plane. Thus, all planes orthogonal to (1, 1, 1) have the form $x + y + z = d$, where d is an arbitrary constant.

Answers to Even-Numbered Exercises

SECTION 1.1

2. (-6, 0, -3)

4. (5, -2)

6. (4, -3)

8. (-2, 5)

10. $\sqrt{13}$

12. $\sqrt{5}$

14. $\sqrt{17}$

16. $5\sqrt{17}$

18. $(2/\sqrt{14}, 1/\sqrt{14}, -3/\sqrt{14})$

20. $(1/\sqrt{5}, -2/\sqrt{5})$

22. (-3, 2, 4)

24. (-3, 5)

26. (3, 2, -1)

28. $4\mathbf{i} + 29\mathbf{j} - 3\mathbf{k}$

32. In $\mathbf{R}^2$, it is a circle of radius 1 centered at the origin. In $\mathbf{R}^3$, it is a sphere of radius 1 centered at the origin.

36. No

40. 50 pounds

SECTION 1.2

2. 12

4. -24

6. -38

8. (-57, -57, 0)

10. 38

12. Acute

14. $5/\sqrt{70}$

16. $1/\sqrt{2}$

18. Of the form $c\,(4, 5, -2)$

20. $3\sqrt{3}$

22. The quantity $\mathbf{v} \cdot \mathbf{w}$ is a scalar and cross product is only defined for two vectors.

24. The quantity $\mathbf{u} \cdot \mathbf{v}$ is a scalar and the symbol $\|\mathbf{x}\|$ is only defined for vectors $\mathbf{x}$.

32. $u_1 = 1/3, \; u_2 = 2/3, \; u_3 = -2/3$

34. 1

38. $500\sqrt{3}$ foot-pounds

40. 75 pound-feet

SECTION 1.3

2. $(-3, 0, 1) \cdot (\mathbf{x} - (1, 2, 3)) = 0$; $-3x + z = 0$

4. $(1, 2, -2) \cdot (\mathbf{x} - (-1, 2, 3)) = 0$; $x + 2y - 2z = -3$

6. $(2, -1, 1) \cdot (\mathbf{x} - (0, 0, 5)) = 0$

8. $(0, 0, 1) \cdot (\mathbf{x} - \mathbf{0}) = 0$

10. $(0, 2) \cdot (\mathbf{x} - (2, -1)) = 0$

12. $(4, 3) \cdot (\mathbf{x} - (1, -3)) = 0$

14. $\mathbf{x}(t) = (3, -1) + t(2, 3)$; $x = 3 + 2t$, $y = -1 + 3t$

16. $\mathbf{x}(t) = (1 - t)(1, 2, -1) + t(2, -1, 3)$; $x = 1 + t$, $y = 2 - 3t$, $z = -1 + 4t$

18. $\mathbf{x}(t) = (1, 2, 3) + t(2, -1, -2)$; $x = 1 + 2t$, $y = 2 - t$, $z = 3 - 2t$

20. $\mathbf{x}(t) = (3, -2) + t(3, -5)$; $x = 3 + 3t$, $y = -2 - 5t$

24. $\mathbf{x}(t) = (0, -5) + t(1, 2)$

28. $(3/2, 3/4, 1)$

34. $\sqrt{33}/3$

36. $3\sqrt{330}/130$

REVIEW EXERCISES

2. $(-2, 4, 12)$

4. $\sqrt{19}$

6. $\sqrt{17}$

8. $(-1/\sqrt{19}, 1/\sqrt{19}, 0, 1/\sqrt{19}, 0, 4/\sqrt{19})$

10. -5

12. $(2/\sqrt{14}, 0, 6/\sqrt{14}, 4/\sqrt{14})$

14. $x + y + z = 1$

16. $2x - y + 3z = -5$

18. $(1, -1, 1, -1)$

20. The expressions in (b) and (f) are not defined.

CHAPTER 2

Euclidean m-*Space and Linear Equations*

Chapter Overview

The topics covered in this chapter are:

1. Definition of $\mathbf{R}^m$ as the set of all ordered *m*-tuples of real numbers; notation for *m-vectors* (elements of $\mathbf{R}^m$) as boldface, lowercase letters, such as **v**. [Section 2.1]
2. *Length* of a vector **v**, $\|\mathbf{v}\|$, in $\mathbf{R}^m$; unit vectors. [2.1]
3. Taking scalar multipies, sums, negatives, and dot products of vectors in $\mathbf{R}^m$; properties of these operations. [2.1]
4. *Orthogonal* vectors. [2.1]
5. *Lines* (two-point and point-parallel forms) and *hyperplanes* (point-normal and standard forms) in $\mathbf{R}^m$. [2.1]
6. Linear equations in *n* variables. [2.1]
7. Representing planes in $\mathbf{R}^m$ in terms of a linear combination of two vectors. [2.1]
8. Terminology for systems of linear equations (*standard form*, *solution*, *consistent*, *inconsistent*). [2.2]
9. Geometric interpretation of the solution as an intersection of hyperplanes. [2.2]
10. Review of the addition method for solving linear systems. [2.2]

11. The *augmented matrix* for a system of linear equations. [2.3]
12. *Elementary row operations* on matrices; *row-equivalent* augmented matrices. [2.3]
13. *Row-reduced echelon form* of a matrix (ones as leading entries, zeros above and below them). [2.3]
14. The Gauss-Jordan elimination method for solving linear systems. [2.3]
15. The Gaussian elimination method for solving linear systems — using reduction to *echelon* form (leading entries are not necessarily ones, zeros only below leading entries). [2.3]
16. Application — Kirchhoff's Laws for an electrical circuit and the corresponding laws for a network of pipes. [2.4]
17. Application — interpolating polynomials. [2.5]

Remarks

The vector notation and operations introduced in Section 2.1will be used throughout the text. If you have covered Chapter 1, Section 2.1 can be dealt with quickly; if you have skipped Chapter 1, some care should be taken with the geometric concepts of line, plane, and hyperplane. (Omitting these concepts is also possible, but they are mentioned from time to time in the text.)

Section 2.2, which reviews some basic concepts of systems of linear equations, is clearly optional. Depending on the students' background and ability, it may be possible to skip it entirely or give it as a reading assignment. Another possibility: after Section 2.1 is covered, proceed directly to Section 2.3, but review the Section 2.2 concepts by solving a simple linear system using both the addition method *and* Gauss-Jordan elimination in parallel columns.

Section 2.3 presents two elimination methods for solving systems of linear equations: *Gauss-Jordan elimination*, which transforms the given augmented matrix to row-reduced echelon form and *Gaussian elimination*, which transforms the augmented matrix to echelon form and then uses back substitution to complete the process. The Gauss-Jordan technique will be used throughout the text; the Gaussian elimination method (given in a *Computational Note*), although more efficient, is optional — it is mentioned again only in Section 3.6, *LU Decomposition*.

Performing the elementary row operations of Section 2.3 can certainly be tedious and time-consuming. If your students have access to a graphing calculator or suitable computer software package, now is the time to introduce them to it.

Sections 2.4 and 2.5 present applications of systems of linear equations. Application sections appear throughout the text; they are presented as soon as the prerequisite linear algebra material has been covered. All these sections are optional.

Solutions to Self-Test

1a. $(2\mathbf{u}) \cdot \mathbf{v} = 2(2, 1, 0, -1) \cdot (-3, 2, -1, 0) = (4, 2, 0, -2) \cdot (-3, 2, -1, 0)$
$= (4)(-3) + (2)(2) + (0)(-1) + (-2)(0) = -8$

1b. $\|\mathbf{v} - \mathbf{u}\| = \|(-3, 2, -1, 0) - (2, 1, 0, -1)\| = \|(-5, 1, -1, 1)\|$
$= \sqrt{(-5)^2 + 1^2 + (-1)^2 + 1^2} = \sqrt{28} = 2\sqrt{7}$

2. *Two-point form*: $\mathbf{x} = (1 - t)\mathbf{p} + t\mathbf{q}$
or $\mathbf{x} = (1 - t)(1, -1, 1, -1) + t(4, 3, 2, 1)$

Point-parallel form: $\mathbf{x} = \mathbf{p} + t(\mathbf{q} - \mathbf{p})$
which yields $\mathbf{x} = (1, -1, 1, -1) + t(3, 4, 1, 2)$

3. A normal to the hyperplane is obtained from the coefficients of the variables: $\mathbf{n} = (1, -1, 2, -1)$. A point on the hyperplane can be obtained by setting x, y, and z equal to 0 in the equation and solving for $w = 6$: (0, 0, 0, 6). Thus, the equation of the hyperplane in point-normal form is:

$$(1, -1, 2, -1) \cdot (\mathbf{x} - (0, 0, 0, 6)) = 0$$

4. Substituting $x = t$, $y = 2t$, and $z - 3t$ into the given equations yields $0 = 0$ and $2t = 1$. Since the second equation is not an identity, the given vector is not a solution of the linear system.

5.
$$\left[\begin{array}{rrr|r} 1 & 1 & -1 & 0 \\ 3 & -2 & 1 & 1 \end{array}\right] \rightarrow \left[\begin{array}{rrr|r} 1 & 1 & -1 & 0 \\ 0 & -5 & 4 & 1 \end{array}\right] \rightarrow$$

$$\left[\begin{array}{rrr|r} 1 & 1 & -1 & 0 \\ 0 & 1 & -4/5 & -1/5 \end{array}\right] \rightarrow \left[\begin{array}{rrr|r} 1 & 0 & -1/5 & 1/5 \\ 0 & 1 & -4/5 & -1/5 \end{array}\right]$$

6. We form the augmented matrix and transform it to row-reduced echelon form:

$$\left[\begin{array}{ccc|c} 1 & 2 & 1 & 1 \\ 1 & 2 & 3 & 5 \\ 0 & 2 & 2 & 2 \end{array}\right] \rightarrow \left[\begin{array}{ccc|c} 1 & 2 & 1 & 1 \\ 0 & 0 & 2 & 4 \\ 0 & 2 & 2 & 2 \end{array}\right] \rightarrow \left[\begin{array}{ccc|c} 1 & 2 & 1 & 1 \\ 0 & 2 & 2 & 2 \\ 0 & 0 & 2 & 4 \end{array}\right] \rightarrow$$

$$\left[\begin{array}{ccc|c} 1 & 2 & 1 & 1 \\ 0 & 1 & 1 & 1 \\ 0 & 0 & 2 & 4 \end{array}\right] \rightarrow \left[\begin{array}{ccc|c} 1 & 0 & -1 & -1 \\ 0 & 1 & 1 & 1 \\ 0 & 0 & 2 & 4 \end{array}\right] \rightarrow \left[\begin{array}{ccc|c} 1 & 0 & -1 & -1 \\ 0 & 1 & 1 & 1 \\ 0 & 0 & 1 & 2 \end{array}\right] \rightarrow$$

$$\left[\begin{array}{ccc|c} 1 & 0 & 0 & 1 \\ 0 & 1 & 0 & -1 \\ 0 & 0 & 1 & 2 \end{array}\right]$$

The last (row-reduced) augmented matrix corresponds to the equations $x = 1$, $y = -1$, $z = 2$. Thus, the solution of the given system is $(1, -1, 2)$.

7. We form the augmented matrix and transform it to row-reduced echelon form:

$$\left[\begin{array}{cccc|c} 1 & -2 & 1 & -1 & 1 \\ 2 & -3 & 0 & 2 & 2 \end{array}\right] \rightarrow \left[\begin{array}{cccc|c} 1 & -2 & 1 & -1 & 1 \\ 0 & 1 & -2 & 4 & 0 \end{array}\right] \rightarrow$$

$$\left[\begin{array}{cccc|c} 1 & 0 & -3 & 7 & 1 \\ 0 & 1 & -2 & 4 & 0 \end{array}\right]$$

The last (row-reduced matrix corresponds to the linear system:

$$\begin{aligned} x_1 \qquad - 3x_3 + 7x_4 &= 1 \\ x_2 - 2x_3 + 4x_4 &= 0 \end{aligned}$$

Solving the first equation for x_1 and the second equation for x_2 yields:

$$\begin{aligned} x_1 &= 1 + 3x_3 - 7x_4 \\ x_2 &= 2x_3 - 4x_4 \end{aligned}$$

Now, setting x_3 and x_4 equal to s and t, respectively, we obtain the two parameter family of solutions: $(1 + 3s - 7t, 2s - 4t, s, t)$

8. To solve the system using Gaussian elimination, we transform the augmented matrix to echelon form:

$$\begin{bmatrix} 1 & 2 & 1 & | & 1 \\ 1 & 2 & 3 & | & 5 \\ 0 & 2 & 2 & | & 2 \end{bmatrix} \rightarrow \begin{bmatrix} 1 & 2 & 1 & | & 1 \\ 0 & 2 & 2 & | & 2 \\ 0 & 0 & 2 & | & 4 \end{bmatrix}$$

The linear system corresponding to the last augmented matrix is:

$$\begin{aligned} x + 2y + z &= 1 \\ 2y + 2z &= 2 \\ 2z &= 4 \end{aligned}$$

We now solve this system by back substitution. The last equation yields $z = 2$, which we substitute into the second equation to get $y = -1$. Then, we substitute $z = 2$ and $y = -1$ into the first equation and obtain $x = 1$. Thus, the solution is $(1, -1, 2)$.

Solutions to Odd-Numbered Review Exercises

1. $$\begin{aligned} \mathbf{x} - 3\mathbf{y} &= (1, 0, 3, 0, 2) - 3(-1, 1, 0, 1, 4) = (1, 0, 3, 0, 2) - (-3, 3, 0, 3, 12) \\ &= (4, -3, 3, -3, -10) \end{aligned}$$

3. $$\begin{aligned} \|-2\mathbf{x}\| &= \|-2(1, 0, 3, 0, 2)\| = \|(-2, 0, -6, 0, -4)\| \\ &= \sqrt{(-2)^2 + 0^2 + (-6)^2 + 0^2 + (-4)^2} = \sqrt{56} = 2\sqrt{14} \end{aligned}$$

5. $$\begin{aligned} \mathbf{x} \cdot \mathbf{y} &= (1, 0, 3, 0, 2) \cdot (-1, 1, 0, 1, 4) \\ &= (1)(-1) + (0)(1) + (3)(0) + (0)(1) + (2)(4) = 7 \end{aligned}$$

7. Let $\mathbf{x} = (x, y, z, u, v)$ and perform the subtraction and dot product operations on the left side as follows:

$$\begin{aligned} (1, 2, 3, 4, 5) \cdot (x - 1, y - 1, z - 1, u - 1, v - 1) &= 0 \\ (1)(x - 1) + (2)(y - 1) + (3)(z - 1) + (4)(u - 1) + (5)(v - 1) &= 0 \\ (x - 1) + (2y - 2) + (3z - 3) + (4u - 4) + (5v - 5) &= 0 \\ x + 2y + 3z + 4u + 5v &= 15 \end{aligned}$$

9. The desired line passes through the point $(0, 0, 0, 0)$ and has the direction $(1, 0, 2, 0)$. Thus, its equation is:

$$\mathbf{x} = (0, 0, 0, 0) + t(1, 0, 2, 0) = t(1, 0, 2, 0)$$

11. Expression (b) is not defined because it is the difference of a scalar and a vector. Expression (c) is not defined because *cross product* is only defined if both vectors are in $\mathbf{R}^3$. Expression (f) is not defined because the dot product of two vectors is a scalar and *norm* is only defined for vectors.

13. For Gauss-Jordan elimination, the augmented matrix on the left is transformed into the row-reduced echelon form on the right:

$$\left[\begin{array}{rr|r} 1 & -1 & -1 \\ -2 & 4 & 2 \end{array}\right] \rightarrow \left[\begin{array}{rr|r} 1 & -2 & -1 \\ 0 & 0 & 0 \end{array}\right]$$

The row-reduced matrix represents the single equation

$$x - 2y = -1$$

or

$$x = -1 + 2y.$$

Thus, the solution is $(-1 + 2s, s)$, where s is a scalar.

For this system, the echelon matrix is the same as the row-reduced echelon matrix. Hence, the solution for Gaussian elimination is identical to the one above.

15. For Gauss-Jordan elimination, the augmented matrix on the left is transformed into the row-reduced echelon form on the right:

$$\left[\begin{array}{rrr|r} 1 & -2 & 1 & 0 \\ -1 & 2 & -1 & 0 \end{array}\right] \rightarrow \left[\begin{array}{rrr|r} 1 & -2 & 1 & 0 \\ 0 & 0 & 0 & 0 \end{array}\right]$$

The row-reduced matrix represents the single equation

$$x_1 - 2x_2 + x_3 = 0$$

or

$$x_1 = 2x_2 - x_3.$$

Thus, the solution is $(2s - t, s, t)$, where s and t are scalars.

For this system, the echelon matrix is the same as the row-reduced echelon matrix. Hence, the solution for Gaussian elimination is identical to the one above.

17. For Gauss-Jordan elimination, the augmented matrix on the left is transformed into the row-reduced echelon form on the right:

$$\left[\begin{array}{ccc|c} 1 & 2 & -3 & 4 \\ 0 & -1 & 1 & -1 \\ 2 & 0 & -1 & 3 \end{array}\right] \rightarrow \left[\begin{array}{ccc|c} 1 & 0 & 0 & 1 \\ 0 & 1 & 0 & 0 \\ 0 & 0 & 1 & -1 \end{array}\right]$$

Consequently, $u = 1$, $v = 0$, and $w = -1$.

Solving by Gaussian elimination, we obtain the following echelon form for the augmented matrix and corresponding system of equations:

$$\left[\begin{array}{ccc|c} 1 & 2 & -3 & 4 \\ 0 & -1 & 1 & -1 \\ 2 & 0 & 9 & -9 \end{array}\right] \qquad \begin{aligned} u + 2v - 3w &= 4 \\ -v + w &= -1 \\ 9w &= -9 \end{aligned}$$

We now solve this system by back-substitution. From the last equation, we obtain $w = -1$. Substituting this value into the second equation yields $v = 0$, and finally substituting these values for v and w into the first equation gives $u = 1$. Thus, the solution of the given system is $(1, 0, -1)$.

19. In both the Gauss-Jordan and Gaussian methods, in transforming the augmented matrix, we obtain a row of zeros to the left of the "vertical bar" and a nonzero number to its right. For Gauss-Jordan elimination, the matrix at this point looks like this:

$$\left[\begin{array}{cccc|c} 1 & 0 & 1 & -1 & 0 \\ 0 & 1 & 1 & 0 & 1 \\ 0 & 0 & 0 & 0 & 3 \end{array}\right]$$

For Gaussian elimination, it looks like this:

$$\left[\begin{array}{cccc|c} 1 & 1 & 2 & -1 & 1 \\ 0 & 3 & 3 & 0 & 3 \\ 0 & 0 & 0 & 0 & 3 \end{array}\right]$$

Consequently, the given system has no solution.

Answers to Even-Numbered Exercises

SECTION 2.1

2. (11, -2, 7, 11)

4. $\sqrt{47}$

6. $2\sqrt{11}$

8. $(1/2\sqrt{11}, -1/2\sqrt{11}, 0, 3/2\sqrt{11})$

10. (-10, 10, 0, -30)

12. 30

16. (1/6, 1/6, 1/2, 0, 5/6)

18. $x_1 - x_3 + x_5 = 0$

20. $\mathbf{x}(t) = (1 - t)(2, 1, 0, 3, 1) + t(1, -1, 3, 0, 5)$; [two-point]
$\mathbf{x}(t) = (2, 1, 0, 3, 1) + t(-1, -2, 3, -3, 4)$; [point-parallel]
$x_1 = 2 - t,\ x_2 = 1 - 2t,\ x_3 = 3t,\ x_4 = 3 - 3t,\ x_5 = 1 + 4t$ [parametric]

22. $(1, 2, -1, 3) \cdot (\mathbf{x} - (-2, 1, 4, 0)) = 0$ [point-normal]
$x_1 + 2x_2 - x_3 + 3x_4 = -4$ [standard]

24. $(2, -3, 0, 1, -1) \cdot (\mathbf{x} - (1, 0, 0, 0, 0)) = 0$

26. It contains a product of variables.

28. It contains $\sin y$.

30. (4, 2, 5/2, 3)

32. -1/12

34. $\mathbf{x} = (3, 1, 0, 2, 1) + t_1(-1, 0, 4, 0, -1) + t_2(-4, 1, 1, 1, 0)$

36. (2, 0, 2, -2)

SECTION 2.2

2. No

4. Yes

6. No

8.
$$\begin{aligned} -x_1 + x_2 + x_3 - x_4 &= 3 \\ x_2 &= 0 \\ x_1 + x_2 + x_3 &= 1 \end{aligned}$$

10.
$$\begin{aligned} x + y &= 0 \\ x - z &= 1 \\ x - y &= 1 \end{aligned}$$

16. $(11/4 + (3/4)t,\ 3/4 - (1/4)t,\ 3/2 + (3/2)t,\ t)$

18. (0, 0)

20. No point of intersection

22. A plane

24. $k = 1$

28.
$$\begin{array}{rcrcrcrcrcrcrcl}
4u_7 & - & u_8 & & & - & u_{12} & & & & & = & 90 \\
-u_7 & + & 4u_8 & - & u_9 & & & - & u_{13} & & & = & 30 \\
 & & -u_8 & + & 4u_9 & & & & & - & u_{14} & = & 140 \\
-u_7 & & & & & + & 4u_{12} & - & u_{13} & & & = & 280 \\
 & & -u_8 & & & - & u_{12} & + & 4u_{13} & - & u_{14} & = & 180 \\
 & & & & -u_9 & & & - & u_{13} & + & 4u_{14} & = & 340
\end{array}$$

SECTION 2.3

2. $\left[\begin{array}{rrrr|r} 1 & 1 & -1 & 0 & 3 \\ 0 & 1 & 1 & -1 & 0 \\ -1 & 0 & 1 & -1 & 1 \end{array}\right]$

4. $\left[\begin{array}{rrrr|r} 1 & 0 & 0 & 1 & 0 \\ 0 & 1 & 1 & 0 & 0 \end{array}\right]$

6. Yes

8. No

10. No

12. $x_1 = -s,\ x_2 = s,\ x_3 = s,$ where s is any scalar

14. $x_1 = 0,\ x_2 = 0,\ x_3 = s,$ where s is any scalar

16. No solution

18. (s, s)

20. $(-1, -2s, 0, s)$

22. No solution

24. $((5/7)s, (1/7)s, -(10/7)s, s)$

26. $(-1 + s, -1/3, 4/3 - s, s)$

38. $\begin{bmatrix} 1 & 0 & 0 \\ 0 & 1 & 0 \\ 0 & 0 & 1 \end{bmatrix}, \begin{bmatrix} 1 & 0 & a \\ 0 & 1 & b \\ 0 & 0 & 1 \end{bmatrix}, \begin{bmatrix} 1 & a & b \\ 0 & 0 & 0 \\ 0 & 0 & 0 \end{bmatrix}, \begin{bmatrix} 1 & a & 0 \\ 0 & 0 & 1 \\ 0 & 0 & 0 \end{bmatrix}, \begin{bmatrix} 0 & 1 & 0 \\ 0 & 0 & 1 \\ 0 & 0 & 0 \end{bmatrix},$

$\begin{bmatrix} 0 & 1 & a \\ 0 & 0 & 0 \\ 0 & 0 & 0 \end{bmatrix}, \begin{bmatrix} 0 & 0 & 1 \\ 0 & 0 & 0 \\ 0 & 0 & 0 \end{bmatrix}, \begin{bmatrix} 0 & 0 & 0 \\ 0 & 0 & 0 \\ 0 & 0 & 0 \end{bmatrix}$

40. **a.** There are 27 equations in 27 unknowns.
b. Each equation contains between three and five unknowns.

SECTION 2.4

2. Current through: 'a' and 'c' is 5/6 amp; 'b' and 'd' is 1/6 amp; 'e' is 1 amp.

4. Current through: 'a' is 0.89 amp; 'b' is 1.11 amp; 'c' is 0.21 amp; 'd' is 0.91 amp; 'e' is 1.13 amp.

6. $x_a = 0.70$, $x_b = 1.27$, $x_c = 0.57$, $x_d = 0.92$, $x_e = 1.49$

8. $x_1 = 48.2$, $x_2 = 229.6$, $x_3 = 51.9$, $x_4 = 170.4$, $x_5 = 3.7$, $x_6 = 59.3$, $x_7 = 55.6$

10. Same answer as Exercise 6.

SECTION 2.5

2. $p(x) = 7/2 - x - (1/2)x^2$

4. $p(x) = 1 - 2x + x^2$

6. $p(x) = -3 - 8x^2 + 2x^4$

8. $p(2.5) = .3999$

REVIEW EXERCISES

2. $\sqrt{19}$

4. $(-1/\sqrt{19}, 1/\sqrt{19}, 0, 1/\sqrt{19}, 4/\sqrt{19})$

6. -5

8. $(3, 1, -2, 1) \cdot (\mathbf{x} - (0, 0, 0, 1)) = 0$

10. $(1, -1, 1, -1)$

12. $\mathbf{x}(t) = (1, 0, 0, 1) + s(-1, 1, 0, 0) + t(-1, 0, 1, 0)$

14. $(3 - s, 4, s)$

16. $(41/29, -32/29, 31/29)$

18. No solution

20. $(-t, s - 2t, s, t)$

CHAPTER 3

Matrices

Chapter Overview

The topics covered in this chapter are:

1. Basic notation and terminology ($m \times n$ *matrix*, *dimensions*, *corresponding entries*, etc.) involving matrices. [Section 3.1]
2. Addition, subtraction, and multiplication of matrices; multiplying a matrix by a scalar, taking the negative of a matrix, finding powers of a matrix. [3.1]
3. Zero and identity matrices. [3.1]
4. The transpose of a matrix. [3.1]
5. Properties of matrix operations. [3.1]
6. The coefficient matrix for a system of linear equations; the matrix form of a linear system, $A\mathbf{x} = \mathbf{b}$. [3.2]
7. The inverse of a matrix; computation by row-reduction. [3.2]
8. Finding the solution of a square, nonsingular linear system by the *method of inverses* ($\mathbf{x} = A^{-1}\mathbf{b}$). [3.2]
9. Application — least square polynomials. [3.3]
10. Application — Leontief models. [3.4]
11. The possible number of solutions for homogenous and nonhomogeneous linear systems. [3.5]

12. The *rank* of a linear system (as the number of nonzero rows in its row-reduced echelon form). [3.5]

13. The relation between the number of solutions (and number of parameters in a solution) of a linear system and the ranks of its coefficient and augmented matrices. [3.5]

14. For an $n \times n$ matrix A, the equivalence of invertibility with: [3.5]

    - rank of $A = n$;
    - uniqueness of solutions of $A\mathbf{x} = \mathbf{b}$, and $A\mathbf{x} = \mathbf{0}$;
    - row-reduced echelon form of A is I.

15. LU decomposition of a matrix as A = LU or, more generally, LU = PA, where P is a *permutation matrix*. [3.6]

16. Using LU decomposition to solve a system of linear equations. [3.6]

17. *Elementary matrices*; relation to elementary row operations and LU decomposition. [3.7]

Remarks

This chapter contains both essential and optional material. Sections 3.1 and 3.2 present fundamental matrix operations, material often taught in high school or College Algebra classes. These sections contain quite a bit of material, but can be covered quickly. Section 3.5, which is more theoretical in nature than any of the prior sections, also contains material that is used throughout the text. The theorems presented here are not difficult to prove and can serve to give the students a taste of "real linear algebra." Note that in addition to the theorems, Section 3.5 also contains the definitions of *homogeneous* linear system and *rank* of a matrix.

The applications sections (3.3 and 3.4) and the material on LU decomposition and elementary matrices (Sections 3.6 and 3.7) are completely optional. (The latter is only used in the proofs of a couple of theorems later in the text.)

Solutions to Self-Test

1. $3A - I = 3\begin{bmatrix} 1 & 0 \\ -1 & 2 \end{bmatrix} - \begin{bmatrix} 1 & 0 \\ 0 & 1 \end{bmatrix} = \begin{bmatrix} 3 & 0 \\ -3 & 6 \end{bmatrix} - \begin{bmatrix} 1 & 0 \\ 0 & 1 \end{bmatrix} = \begin{bmatrix} 2 & 0 \\ -3 & 5 \end{bmatrix}$

2. $$AB = \begin{bmatrix} 1 & 0 \\ -1 & 2 \end{bmatrix} \begin{bmatrix} 2 & 0 & 1 \\ 3 & -1 & 0 \end{bmatrix}$$

$$= \begin{bmatrix} (1)(2)+(0)(3) & (1)(0)+(0)(-1) & (1)(1)+(0)(0) \\ (-1)(2)+(2)(3) & (-1)(0)+(2)(-1) & (-1)(1)+(2)(0) \end{bmatrix}$$

$$= \begin{bmatrix} 2 & 0 & 1 \\ 4 & -2 & -1 \end{bmatrix}$$

3. B^T is the 3×2 matrix whose first column is the first row of B and whose second column is the second row of B:

$$\begin{bmatrix} 2 & 3 \\ 0 & -1 \\ 1 & 0 \end{bmatrix}$$

4. Augment the given matrix C by the third order identity matrix I and transform the augmented matrix to row-reduced echelon form:

$$\left[\begin{array}{ccc|ccc} 1 & 2 & 3 & 1 & 0 & 0 \\ 0 & 1 & 4 & 0 & 1 & 0 \\ 0 & 0 & 1 & 0 & 0 & 1 \end{array}\right] \rightarrow \left[\begin{array}{ccc|ccc} 1 & 0 & 0 & 1 & -2 & 5 \\ 0 & 1 & 0 & 0 & 1 & -4 \\ 0 & 0 & 1 & 0 & 0 & 1 \end{array}\right]$$

Then, C^{-1} is the matrix whose columns are the last three columns of the row-reduced matrix:

$$\begin{bmatrix} 1 & -2 & 5 \\ 0 & 1 & -4 \\ 0 & 0 & 1 \end{bmatrix}$$

5. The given system has the form $A\mathbf{x} = \mathbf{b}$, where

$$A = \begin{bmatrix} 2 & -1 \\ -1 & 1 \end{bmatrix}, \quad \mathbf{x} = \begin{bmatrix} x \\ y \end{bmatrix}, \quad \mathbf{b} = \begin{bmatrix} -1 \\ 2 \end{bmatrix}$$

To solve this system by the method of inverses, we find A^{-1} and then multiply it by **b**. The result yields the solution **x**. Now,

$$A^{-1} = \begin{bmatrix} 1 & 1 \\ 1 & 2 \end{bmatrix} \quad \text{and} \quad A^{-1}\mathbf{b} = \begin{bmatrix} 1 \\ 3 \end{bmatrix}$$

Thus, the solution of the given system is (1, 3).

6. Notice that the both A and $[A|\mathbf{b}]$ are in row-reduced echelon form.

a. Since the row-reduced echelon form of A has two nonzero rows, its rank is 2.

b. Since the ranks of A and $[A|\mathbf{b}]$ are the same, and this number (2) is one less than the number of unknowns, the given system has a one-parameter family of (infinitely many) solutions.

c. Since the order of A is 3 but its rank is 2, A is not invertible.

7. a. A linear system might not have any solution, so the statement is false. (A *homogeneous* linear system must have at least one solution.)

b. This statement is true due to Theorem 5 of Section 3.5.

8. Let matrices O and I be as follows

$$O = \begin{bmatrix} 0 & 0 & 0 \\ 0 & 0 & 0 \\ 0 & 0 & 0 \end{bmatrix}, \quad I = \begin{bmatrix} 1 & 0 & 0 \\ 0 & 1 & 0 \\ 0 & 0 & 1 \end{bmatrix}$$

and perform Gaussian elimination on the given matrix A while recording the corresponding adjustments to O and I:

$$A \rightarrow \begin{bmatrix} 1 & 2 & 3 \\ 0 & 0 & 4 \\ 0 & 2 & -6 \end{bmatrix}, \quad O \rightarrow \begin{bmatrix} 0 & 0 & 0 \\ -1 & 0 & 0 \\ 2 & 0 & 0 \end{bmatrix}, \quad I \rightarrow \begin{bmatrix} 1 & 0 & 0 \\ 0 & 1 & 0 \\ 0 & 0 & 1 \end{bmatrix}$$

$$A \rightarrow \begin{bmatrix} 1 & 2 & 3 \\ 0 & 2 & -6 \\ 0 & 0 & 4 \end{bmatrix}, \quad O \rightarrow \begin{bmatrix} 0 & 0 & 0 \\ 2 & 0 & 0 \\ -1 & 0 & 0 \end{bmatrix}, \quad I \rightarrow \begin{bmatrix} 1 & 0 & 0 \\ 0 & 0 & 1 \\ 0 & 1 & 0 \end{bmatrix}$$

Making the appropriate adjustment to the main diagonal of the second matrix, we have

$$U = \begin{bmatrix} 1 & 2 & 3 \\ 0 & 2 & -6 \\ 0 & 0 & 4 \end{bmatrix}, \quad L = \begin{bmatrix} 1 & 0 & 0 \\ 2 & 1 & 0 \\ -1 & 0 & 1 \end{bmatrix}, \quad P = \begin{bmatrix} 1 & 0 & 0 \\ 0 & 0 & 1 \\ 0 & 1 & 0 \end{bmatrix}.$$

9. With L, U, and P as in the solution of Problem **8**, we solve $(PA)\mathbf{x} = P\mathbf{b}$, by first solving $L\mathbf{y} = P\mathbf{b}$ by forward substitution:

$$\begin{aligned} y_1 \qquad\qquad &= 1 \\ 2y_1 + y_2 \qquad &= 2 \\ -y_1 \qquad + y_3 &= 3 \end{aligned}$$

Doing so, we obtain $\mathbf{y} = (1, 0, 4)$. Then, we solve $U\mathbf{x} = \mathbf{y}$

$$\begin{aligned} x_1 + 2x_2 + 3x_3 &= 1 \\ 2x_2 - 6x_3 &= 0 \\ 4x_3 &= 4 \end{aligned}$$

by forward substitution. Doing so, we obtain $\mathbf{x} = (-8, 3, 1)$.

10. To obtain the required elementary matrices, we first note each of the row operations used in the solution of Problem 8 and record the corresponding elementary matrix. Using the notation of Theorem 1 of Section 3.7, we have $E_1 = E_{21}(1)$, $E_2 = E_{31}(-2)$, and $E_3 = P_{23}$. Then, to obtain leading ones in each nonzero row, we use $E_4 = E_{22}(1/2)$ and $E_5 = E_{33}(1/4)$. Finally, to obtain zeros above the leading ones, we use $E_6 = E_{12}(-2)$, $E_7 = E_{13}(-9)$, and $E_8 = E_{23}(3)$.

Solutions to Odd-Numbered Review Exercises

1. $$2A - 3B = 2\begin{bmatrix} 1 & 0 & 1 \\ 0 & 1 & 3 \\ 0 & 0 & 3 \end{bmatrix} - 3\begin{bmatrix} -1 & 1 & 1 \\ 1 & 3 & 1 \\ 2 & 0 & -1 \end{bmatrix} = \begin{bmatrix} 2 & 0 & 2 \\ 0 & 2 & 6 \\ 0 & 0 & 6 \end{bmatrix} - \begin{bmatrix} -3 & 3 & 3 \\ 3 & 9 & 3 \\ 6 & 0 & -3 \end{bmatrix}$$

$$= \begin{bmatrix} 5 & -3 & -1 \\ -3 & -7 & 3 \\ -6 & 0 & 9 \end{bmatrix}$$

3. $$AB = \begin{bmatrix} 1 & 0 & 1 \\ 0 & 1 & 3 \\ 0 & 0 & 3 \end{bmatrix} \begin{bmatrix} -1 & 1 & 1 \\ 1 & 3 & 1 \\ 2 & 0 & -1 \end{bmatrix}$$

$$= \begin{bmatrix} (1)(-1)+(0)(1)+(1)(2) & (1)(1)+(0)(3)+(1)(0) & (1)(1)+(0)(1)+(1)(-1) \\ (0)(-1)+(1)(1)+(3)(2) & (0)(1)+(1)(3)+(3)(0) & (0)(1)+(1)(1)+(3)(-1) \\ (0)(-1)+(0)(1)+(3)(2) & (0)(1)+(0)(3)+(3)(0) & (0)(1)+(0)(1)+(3)(-1) \end{bmatrix}$$

$$= \begin{bmatrix} 1 & 1 & 0 \\ 7 & 3 & -2 \\ 6 & 0 & -3 \end{bmatrix}$$

5. We transform the augmented matrix

$$\left[\begin{array}{ccc|ccc} 1 & 0 & 1 & 1 & 0 & 0 \\ 0 & 1 & 3 & 0 & 1 & 0 \\ 0 & 0 & 3 & 0 & 0 & 1 \end{array}\right]$$

into row-reduced echelon form:

$$\left[\begin{array}{ccc|ccc} 1 & 0 & 0 & 1 & 0 & -1/3 \\ 0 & 1 & 0 & 0 & 1 & -1 \\ 0 & 0 & 1 & 0 & 0 & 1/3 \end{array}\right]$$

Thus, the inverse of A is:

$$\begin{bmatrix} 1 & 0 & -1/3 \\ 0 & 1 & -1 \\ 0 & 0 & 1/3 \end{bmatrix}$$

7. $$A^TB = \begin{bmatrix} 1 & 0 & 0 \\ 0 & 1 & 0 \\ 1 & 3 & 3 \end{bmatrix} \begin{bmatrix} -1 & 1 & 1 \\ 1 & 3 & 1 \\ 2 & 0 & -1 \end{bmatrix} = \begin{bmatrix} -1 & 1 & 1 \\ 1 & 3 & 1 \\ 8 & 10 & 1 \end{bmatrix}$$

We now attempt to find the inverse of this last matrix by transforming $[A^TB|I]$ into row-reduced echelon form. In doing so, we obtain a row of zeros to the left of the "vertical bar," showing that the inverse of A^TB dies not exist.

9. The row-reduced echelon form of A is the identity matrix, which has three nonzero rows. Hence, the rank of A is 3.

11. $$-A^2 = -\begin{bmatrix} 1 & 0 & 1 \\ 0 & 1 & 3 \\ 0 & 0 & 3 \end{bmatrix}\begin{bmatrix} 1 & 0 & 1 \\ 0 & 1 & 3 \\ 0 & 0 & 3 \end{bmatrix} = -\begin{bmatrix} 1 & 0 & 4 \\ 0 & 1 & 12 \\ 0 & 0 & 9 \end{bmatrix} = \begin{bmatrix} -1 & 0 & -4 \\ 0 & -1 & -12 \\ 0 & 0 & -9 \end{bmatrix}$$

13. The matrix form for this linear system is $A\mathbf{x} = \mathbf{b}$, where

$$A = \begin{bmatrix} 1 & -2 \\ -2 & 4 \end{bmatrix}, \quad \mathbf{x} = \begin{bmatrix} x \\ y \end{bmatrix}, \quad \mathbf{b} = \begin{bmatrix} -1 \\ 2 \end{bmatrix}$$

The method of inverses does not work because A is not invertible.

15. The matrix form for this linear system is $A\mathbf{x} = \mathbf{b}$, where

$$A = \begin{bmatrix} 1 & -2 & 1 \\ -1 & 2 & -1 \end{bmatrix}, \quad \mathbf{x} = \begin{bmatrix} x_1 \\ x_2 \\ x_3 \end{bmatrix}, \quad \mathbf{b} = \begin{bmatrix} 0 \\ 0 \end{bmatrix}$$

The method of inverses does not work because A is not a square matrix.

17. The matrix form for this linear system is $A\mathbf{x} = \mathbf{b}$, where

$$A = \begin{bmatrix} 1 & 2 & -3 \\ 0 & -1 & 1 \\ 2 & 0 & -1 \end{bmatrix}, \quad \mathbf{x} = \begin{bmatrix} u \\ v \\ w \end{bmatrix}, \quad \mathbf{b} = \begin{bmatrix} 4 \\ -1 \\ 3 \end{bmatrix}$$

To solve the system by the method of inverses, we first compute A^{-1} and then multiply it by $\mathbf{b}$:

$$A^{-1}\mathbf{b} = \begin{bmatrix} -1 & -2 & 1 \\ -2 & -5 & 1 \\ -2 & -4 & 1 \end{bmatrix} \begin{bmatrix} 4 \\ -1 \\ 3 \end{bmatrix} = \begin{bmatrix} 1 \\ 0 \\ -1 \end{bmatrix}$$

Thus, the solution is $(1, 0, -1)$.

19. The coefficient matrix is in row-reduced echelon form and has three nonzero rows. Hence, its rank is three, which is the same as the number of unknowns. Thus, there is either a unique solution or no solution. But, the system is homogeneous, so there must be a (unique) solution, $\mathbf{0}$.

21. Row-reducing shows that the augmented matrix has rank two, which is also the rank of the coefficient matrix. Thus, there are infinitely many solutions, and since the system has three unknowns, these solutions form a one- (= 3 - 2) parameter family.

23. To confirm this fact, just multiply the given matrices L and U and see that the result is indeed B.

25. Letting A (the coefficient matrix), O, and I be as follows

$$A = \begin{bmatrix} 1 & -2 \\ -2 & 4 \end{bmatrix}, \quad O = \begin{bmatrix} 0 & 0 \\ 0 & 0 \end{bmatrix}, \quad I = \begin{bmatrix} 1 & 0 \\ 0 & 1 \end{bmatrix}$$

we perform Gaussian elimination on A while recording the adjustments to O and I:

$$A \rightarrow \begin{bmatrix} 1 & -2 \\ 0 & 0 \end{bmatrix}, \quad O \rightarrow \begin{bmatrix} 0 & 0 \\ -2 & 0 \end{bmatrix}, \quad I \rightarrow \begin{bmatrix} 1 & 0 \\ 0 & 1 \end{bmatrix}$$

Making the appropriate adjustments to the main diagonal of the second matrix, we have

$$U = \begin{bmatrix} 1 & -2 \\ 0 & 0 \end{bmatrix}, \quad L = \begin{bmatrix} 1 & 0 \\ -2 & 1 \end{bmatrix}, \quad P = \begin{bmatrix} 1 & 0 \\ 0 & 1 \end{bmatrix}.$$

We now solve $(PA)\mathbf{x} = (LU)\mathbf{x} = P\mathbf{b}$ by first solving $L\mathbf{y} = P\mathbf{b}$ by forward substitution:

$$\begin{aligned} y_1 \quad\quad &= -1 \\ -2y_1 + y_2 &= 2 \end{aligned}$$

Doing so, we obtain the solution $\mathbf{y} = (-1, 0)$. Finally, we solve

$$x - 2y = -1 \quad \text{for} \quad x = 2y - 1.$$

Thus, there are infinitely many solutions of the form $(2t - 1, t)$, where t is an arbitrary real number.

27. Letting A (the coefficient matrix), O, and I be as follows

$$A = \begin{bmatrix} 1 & -2 & 1 \\ -1 & 2 & -1 \end{bmatrix}, \quad O = \begin{bmatrix} 0 & 0 \\ 0 & 0 \end{bmatrix}, \quad I = \begin{bmatrix} 1 & 0 \\ 0 & 1 \end{bmatrix}$$

we perform Gaussian elimination on A while recording the adjustments to O and I:

$$A \rightarrow \begin{bmatrix} 1 & -2 & 1 \\ 0 & 0 & 0 \end{bmatrix}, \quad O \rightarrow \begin{bmatrix} 0 & 0 \\ -1 & 0 \end{bmatrix}, \quad I \rightarrow \begin{bmatrix} 1 & 0 \\ 0 & 1 \end{bmatrix}$$

Making the appropriate adjustments to the main diagonal of the second matrix, we have

$$U = \begin{bmatrix} 1 & -2 & 1 \\ 0 & 0 & 0 \end{bmatrix}, \quad L = \begin{bmatrix} 1 & 0 \\ 1 & 1 \end{bmatrix}, \quad P = \begin{bmatrix} 1 & 0 \\ 0 & 1 \end{bmatrix}.$$

We now solve $(PA)\mathbf{x} = (LU)\mathbf{x} = P\mathbf{b}$ by first solving $L\mathbf{y} = P\mathbf{b}$ by forward substitution:

$$\begin{aligned} y_1 \quad\quad &= 0 \\ -y_1 + y_2 &= 0 \end{aligned}$$

Doing so, we obtain the solution $\mathbf{y} = (0, 0)$. (*Note*: When the given system is homogeneous, as is the case here, this part of the exercise is unnecessary; $L\mathbf{x} = P\mathbf{0}$ always has the unique solution, $\mathbf{x} = \mathbf{0}$.) Finally, we solve

$$x_1 - 2x_2 + x_3 = 0 \quad \text{for} \quad x_1 = 2x_2 - x_3.$$

Thus, there are infinitely many solutions of the form $(2s - t, s, t)$, where s and t are arbitrary real numbers.

29. Letting A (the coefficient matrix), O, and I be as follows

$$A = \begin{bmatrix} 1 & 2 & -3 \\ 0 & -1 & 1 \\ 2 & 0 & -1 \end{bmatrix}, \quad O = \begin{bmatrix} 0 & 0 & 0 \\ 0 & 0 & 0 \\ 0 & 0 & 0 \end{bmatrix}, \quad I = \begin{bmatrix} 1 & 0 & 0 \\ 0 & 1 & 0 \\ 0 & 0 & 1 \end{bmatrix}$$

we perform Gaussian elimination on A while recording the adjustments to O and I:

$$A \rightarrow \begin{bmatrix} 1 & 2 & -3 \\ 0 & -1 & 1 \\ 0 & -4 & 5 \end{bmatrix}, \quad O \rightarrow \begin{bmatrix} 0 & 0 & 0 \\ 0 & 0 & 0 \\ 2 & 0 & 0 \end{bmatrix}, \quad I \rightarrow \begin{bmatrix} 1 & 0 & 0 \\ 0 & 1 & 0 \\ 0 & 0 & 1 \end{bmatrix}$$

$$A \rightarrow \begin{bmatrix} 1 & 2 & -3 \\ 0 & -1 & 1 \\ 0 & 0 & 1 \end{bmatrix}, \quad O \rightarrow \begin{bmatrix} 0 & 0 & 0 \\ 0 & 0 & 0 \\ 2 & 4 & 0 \end{bmatrix}, \quad I \rightarrow \begin{bmatrix} 1 & 0 & 0 \\ 0 & 1 & 0 \\ 0 & 0 & 1 \end{bmatrix}$$

Making the appropriate adjustments to the main diagonal of the second matrix, we have

$$U = \begin{bmatrix} 1 & 2 & -3 \\ 0 & -1 & 1 \\ 0 & 0 & 1 \end{bmatrix}, \quad L = \begin{bmatrix} 1 & 0 & 0 \\ 0 & 1 & 0 \\ 2 & 4 & 1 \end{bmatrix}, \quad P = \begin{bmatrix} 1 & 0 & 0 \\ 0 & 1 & 0 \\ 0 & 0 & 1 \end{bmatrix}$$

We now solve $(PA)\mathbf{x} = (LU)\mathbf{x} = P\mathbf{b}$ by first solving $L\mathbf{y} = P\mathbf{b}$ by forward substitution:

$$\begin{aligned} y_1 \qquad\qquad &= 4 \\ y_2 \qquad &= -1 \\ 2y_1 + 4y_2 + y_3 &= 3 \end{aligned}$$

Doing so, we obtain the solution $\mathbf{y} = (4, -1, -1)$. Finally, we solve

$$\begin{aligned} x_1 + 2x_2 - 3x_3 &= 4 \\ -x_2 + x_3 &= -1 \\ x_3 &= -1 \end{aligned}$$

by backward substitution, to obtain $\mathbf{x} = (1, 0, -1)$.

Answers to Even-Numbered Exercises

SECTION 3.1

2. $\begin{bmatrix} -1 & -4 & 3 \\ -1 & -2 & 1 \\ 1 & 1 & -3 \end{bmatrix}$

4. $\begin{bmatrix} 0 & 0 & 1 \\ 2 & 0 & 3 \\ 0 & 1 & -1 \end{bmatrix}$

6. $\begin{bmatrix} -1 & 4\lambda & 1-2\lambda \\ 2+3\lambda & -1+\lambda & 3+2\lambda \\ -\lambda & 1 & -2+\lambda \end{bmatrix}$

8. $\begin{bmatrix} 1 & 0 & 0 \\ 0 & 1 & 0 \\ 0 & 0 & 1 \end{bmatrix}$

10. $\begin{bmatrix} -1 & -6 & 5 \\ -4 & -7 & 1 \\ 3 & -3 & 0 \end{bmatrix}$

12. $\begin{bmatrix} 8 & -1 & 1 \\ -6 & 1 & 1 \\ 16 & 2 & -3 \end{bmatrix}$

14. The matrix A

16. $\mathbf{a}^1 \cdot \mathbf{a}^3 = -1, \quad \mathbf{a}_1 \cdot \mathbf{a}_2 = 1$

18. $AB = \begin{bmatrix} -3 & 2 & 1 \\ 0 & 0 & 0 \\ 0 & 0 & 0 \end{bmatrix}; \quad BA = \begin{bmatrix} -3 \end{bmatrix}$

20. $AB = \begin{bmatrix} 3 & 2 & 1 \\ 14 & 36 & -22 \end{bmatrix};$ BA is not defined.

22. AB is not defined; BA is the 4×3 zero matrix.

24. $\begin{bmatrix} 15/2 & 6 \\ 15 & 27/4 \end{bmatrix}$

26. $A^2 = \begin{bmatrix} 7 & 10 \\ 15 & 22 \end{bmatrix}, \quad A^3 = \begin{bmatrix} 37 & 54 \\ 81 & 118 \end{bmatrix}$

28. $A^2 = \begin{bmatrix} 1 & 0 & 6 \\ 1 & 1 & -4 \\ 0 & 0 & 4 \end{bmatrix}, \quad A^3 = \begin{bmatrix} 1 & 0 & 14 \\ 2 & -1 & -8 \\ 0 & 0 & 8 \end{bmatrix}$

36. $2^{n-1}\begin{bmatrix} 1 & -1 \\ -1 & 1 \end{bmatrix}$

38. For example, $\begin{bmatrix} 1 & 0 \\ 0 & 0 \end{bmatrix}$

42. **a.** The number of paths of length two from node 1:
to node 1 is one;
to node 2 is zero;
to node 3 is zero;
to node 4 is two.

b. $M^2 = \begin{bmatrix} 1 & 0 & 0 & 2 \\ 0 & 1 & 1 & 0 \\ 0 & 0 & 0 & 0 \\ 0 & 0 & 0 & 0 \end{bmatrix}$

The entry in the i,j-position in the matrix M^2 is equal to the number of paths of length two from node i to node j.

c. $M^3 = \begin{bmatrix} 0 & 1 & 1 & 0 \\ 1 & 0 & 0 & 1 \\ 0 & 0 & 0 & 0 \\ 0 & 0 & 0 & 0 \end{bmatrix}$

The entry in the i,j-position in the matrix M^3 is equal to the number of paths of length three from node i to node j. There is one path of length three from node 2 to node 4.

d. $M^2 + M^3 = \begin{bmatrix} 1 & 1 & 1 & 2 \\ 1 & 1 & 1 & 1 \\ 0 & 0 & 0 & 0 \\ 0 & 0 & 0 & 0 \end{bmatrix}$

The entry in the i,j-position in the matrix $M^2 + M^3$ is equal to the number of paths of lengths two or three from node i to node j.

SECTION 3.2

2. $\begin{bmatrix} 2 & -3 & 0 & 1 \\ 0 & 1 & -1 & 3 \\ 1 & 0 & 0 & 1 \end{bmatrix} \begin{bmatrix} x_1 \\ x_2 \\ x_3 \\ x_4 \end{bmatrix} = \begin{bmatrix} 6 \\ 0 \\ 0 \end{bmatrix}$

4. $\begin{bmatrix} 1 & 0 \\ 1 & 1 \\ 1 & 1 \end{bmatrix} \begin{bmatrix} x_1 \\ x_2 \end{bmatrix} = \begin{bmatrix} 2 \\ 3 \\ 3 \end{bmatrix}$

10. $\begin{bmatrix} 2 & 1 \\ 3/2 & 3/2 \end{bmatrix}$

12. Not invertible

14. $\begin{bmatrix} 1 & 0 & 0 \\ 1/2 & 1/2 & 0 \\ 5/2 & 1/2 & 1 \end{bmatrix}$

16. $\frac{1}{2}\begin{bmatrix} 1 & 1 & 1 & 1 \\ -1 & 1 & 1 & 1 \\ 1 & -1 & 1 & 1 \\ 1 & -1 & -1 & 1 \end{bmatrix}$

18. $c = -3/2$

20. (1, -1)

22. (12, 10, -17, -1)

24. (9/10, -7/10)

26. (-1, 0, 0)

28. No inverse

30. $\begin{bmatrix} 2/7 & 1/7 \\ -1/14 & 3/14 \end{bmatrix}$

SECTION 3.3

2. $p(x) = -22 + (23.8)x$

4. $p(x) = -7/15 + (1/10)x + (11/15)x^2$

6.

$$\begin{aligned} 7b_1 + 21b_2 + 91b_3 + 441b_4 + 2275b_5 &= 18 \\ 21b_1 + 91b_2 + 441b_3 + 2275b_4 + 12201b_5 &= 81 \\ 91b_1 + 441b_2 + 2275b_3 + 12201b_4 + 67171b_5 &= 391 \\ 441b_1 + 2275b_2 + 12201b_3 + 67171b_4 + 376761b_5 &= 1989 \\ 2275b_1 + 12201b_2 + 67171b_3 + 376761b_4 + 2142595b_5 &= 10519 \end{aligned}$$

8. $p(x) = 173 + 62.6x - 37.1x^2$

SECTION 3.4

2. A: $3s$; B: $12s$; C: $15s$ for $s > 0$

4. A should price its production to yield 50% of the total revenue; B, 25%; and C, 25%.

6. A: 38.65; B: 38.54; C: 14.13

SECTION 3.5

2. Nonhomogeneous

4. 2

6. 3

8. **a.** $ad = bc$, but $af \neq ce$

b. $ad \neq bc$

c. $ad = bc$ and $af = ce$

10. False

12. True

14. True

16. No solutions or infinitely many solutions

18. Zero, one, or infinitely many solutions

SECTION 3.6

6. No permutations were necessary.

8. $L = \begin{bmatrix} 1 & 0 & 0 \\ 2 & 1 & 0 \\ 1 & 0 & 1 \end{bmatrix}$, $U = \begin{bmatrix} 3 & 1 & 2 \\ 0 & -1 & -1 \\ 0 & 0 & -4 \end{bmatrix}$, $P = \begin{bmatrix} 1 & 0 & 0 \\ 0 & 0 & 1 \\ 0 & 1 & 0 \end{bmatrix}$

10. $L = \begin{bmatrix} 1 & 0 & 0 \\ 2 & 1 & 0 \\ 3 & 4/5 & 1 \end{bmatrix}$, $U = \begin{bmatrix} 1 & 2 & -1 & 1 \\ 0 & -5 & 4 & -1 \\ 0 & 0 & 4/5 & -11/5 \end{bmatrix}$, $P = I$

12. $\mathbf{x} = (13/6, 4/3)$

14. $\mathbf{x} = (15/4, -5. 1/2)$

16. $\mathbf{x} = (3/4, -5/6, 13/12. 17/12)$

18. $\mathbf{x}(t) = (3/2 - (3/2)t, 1/2 + (1/2)t, -1/2 + (3/2)t, t)$

22. (a)

(b) For each graph, the in-degree is 1 and the out-degree is 1.

SECTION 3.7

2. Multiplies the first row (of a matrix with three rows) by 2.

4. $\begin{bmatrix} 1 & 0 & 0 \\ 0 & 0 & 1 \\ 0 & 1 & 0 \end{bmatrix}$

6. $\begin{bmatrix} 1 & -3 & 0 \\ 0 & 1 & 0 \\ 0 & 0 & 1 \end{bmatrix}$

12. The results are analogous to those of Theorem 1, but with respect to columns instead of rows. For example, postmultiplication by $E_{ij}(c)$ adds c times column i to column j.

14. At this point, all that need be done is to add the appropriate multiple of the second row to the third row. This requires one division to obtain the row multiplier plus one multiplication and one addition to create the 0 in the 3,2-position of the matrix; a total of 3 operations.

16. To obtain row multipliers for the first $n - 1$ columns requires $(n - 1) + (n - 2) + \ldots + 2 + 1$ divisions. To obtain zeros below the diagonal entries in the first $n - 1$ columns requires $2[(n - 1)^2 + (n - 2)^2 + \ldots + 2^2 + 1^2]$ multiplications/additions. The total number of operations is the sum of these two expressions.

REVIEW EXERCISES

2. $\begin{bmatrix} -2 & 1 & 3 \\ 1 & 2 & 2 \\ 2 & 0 & 2 \end{bmatrix}$

4. $\begin{bmatrix} -1 & 1 & 5 \\ 1 & 3 & 13 \\ 2 & 0 & -1 \end{bmatrix}$

6. B^{-1} does not exist.

8. $\frac{1}{9}\begin{bmatrix} 10 & 3 & -1 \\ 3 & 18 & -3 \\ -1 & -3 & 1 \end{bmatrix}$

10. 2

12. $\begin{bmatrix} 1 & 0 & 13 \\ 0 & 1 & 39 \\ 0 & 0 & 27 \end{bmatrix}$

14. The method of inverses cannot be used; the system is not square.

16. (41/29, -32/29, 31/29)

18. The method of inverses cannot be used; the inverse of the coefficient matrix does not exist.

20. There are infinitely many solutions; two parameters.

22. There is no solution.

24. A factorization that avoids fractions is LU = PA, where

$$L = \begin{bmatrix} 1 & 0 & 0 \\ -2 & 1 & 0 \\ -1 & 2 & 1 \end{bmatrix}, \quad U = \begin{bmatrix} -1 & 1 & 1 \\ 0 & 2 & 1 \\ 0 & 0 & 0 \end{bmatrix}, \quad P = \begin{bmatrix} 1 & 0 & 0 \\ 0 & 0 & 1 \\ 0 & 1 & 0 \end{bmatrix}$$

26. (3 - s, 4, s)

28. (41/29, -32/29, 31/29)

30. No solution

32. **i.** The inverse, $\begin{bmatrix} 1 & -2 & 0 \\ 0 & 1 & 0 \\ 0 & 0 & 1 \end{bmatrix}$, adds -2 times row 2 to row 1.

ii. The inverse, $\begin{bmatrix} 0 & 0 & 1 \\ 0 & 1 & 0 \\ 1 & 0 & 0 \end{bmatrix}$, interchanges rows 1 and 3.

iii. The inverse, $\begin{bmatrix} 1 & 0 & 0 \\ 0 & 1 & 0 \\ 0 & 0 & 1/2 \end{bmatrix}$, multiplies row 3 by 1/2.

CHAPTER 4

Determinants

Chapter Overview

The topics covered in this chapter are:

1. Definition of *determinant* in terms of cofactor expansion. (*Submatrices*, *cofactors*, and *minors* are defined first.) [Section 4.1]
2. *Lower triangular*, *upper triangular*, and *diagonal* matrices. [4.1]
3. The effect that performing each of the three elementary row operations has on the determinant of a matrix. [4.2]
4. Basic theorems concerning determinants:
   - If all entries in any row or column of a square matrix are 0, then its determinant is 0. [4.1]
   - The determinant of an upper triangular, lower triangular, or diagonal matrix is equal to the product of its diagonal entries. [4.1]
   - A system of n linear equations in n unknowns has a unique solution if and only if the determinant of its coefficient matrix is nonzero. [4.2]
   - If A and B are square matrices of the same order, then $\det(AB) = \det(A) \cdot \det(B)$. [4.2]
   - If A is a square matrix, then $\det(A^T) = \det(A)$. [4.2]
5. Cramer's rule for solving a square linear system. [4.3]
6. The adjoint of a square matrix A; using the adjoint to find A^{-1}. [4.3]

Remarks

Determinants are defined in this text using a recursive definition involving cofactor expansion along the first row. We have found this approach to be much easier for the student (although certainly less elegant) than a more formal treatment of determinants (for example, one using permutations). Unfortunately, our definition makes it much more difficult to prove that cofactor expansion along *any* row or column yields the same determinant, and we have omitted this proof.

All the material in Sections 4.1 and 4.2 (the definition and basic properties of the determinant) is used later in the text. Section 4.3 (which deals with Cramer's rule and the adjoint form of the inverse), however, is optional and can be skipped without creating any future problems. If Cramer's rule is covered, you should probably emphasize that, although it provides a closed form solution of a linear system, it has major deficiencies: The row-reduction techniques of Section 2.3 are more efficient and more informative (in the case where the determinant is 0).

Solutions to Self-Test

1. There is no "best" way to do this one but, in general, it is easier to subtract multiples of a row or column from others to simplify the arithmetic. We subtract 3 times the second row from the first and 2 times the second row from the third:

$$\det\begin{bmatrix} 3 & 2 & 1 \\ 1 & 4 & 2 \\ 2 & 5 & 3 \end{bmatrix} = \det\begin{bmatrix} 0 & -10 & -5 \\ 1 & 4 & 2 \\ 0 & -3 & -1 \end{bmatrix} = -\det\begin{bmatrix} -10 & -5 \\ -3 & -1 \end{bmatrix} = 5$$

2. To eliminate the decimals we can factor 0.01 out of each row (or column) to obtain a matrix with small integer coefficients:

$$(.01)^3\det\begin{bmatrix} 2 & 3 & 1 \\ -1 & 6 & 2 \\ 2 & -9 & -3 \end{bmatrix} = (.01)^3(0) = 0$$

3. Interchanging the first and second rows and likewise the third and fourth, we introduce two factors of -1 or no change in the value of the determinant. Since the resulting matrix is a diagonal, its determinant is $(2)(3)(3)(2) = 36$.

4. The matrix is upper triangular, so the given determinant is the product of the diagonal entries. Thus, the equation simplifies to

$$(1 - t)(2 - t)(3 - t) = 0$$

and its solutions are $t = 1, 2, 3$.

5. The determinant of the coefficient matrix is 7, so:

$$x = \frac{\det\begin{bmatrix} 2 & -1 \\ 3 & 4 \end{bmatrix}}{7} = \frac{11}{7}, \quad y = \frac{\det\begin{bmatrix} 1 & 2 \\ 3 & 3 \end{bmatrix}}{7} = -\frac{3}{7}$$

6. The determinant of the coefficient matrix is 15, so:

$$x = \frac{\det\begin{bmatrix} 3 & 1 & 1 \\ 2 & 0 & -1 \\ 1 & 5 & 2 \end{bmatrix}}{15} = \frac{20}{15} = \frac{4}{3}$$

$$y = \frac{\det\begin{bmatrix} 2 & 3 & 1 \\ 3 & 2 & -1 \\ 4 & 1 & 2 \end{bmatrix}}{15} = -\frac{25}{15} = -\frac{5}{3}$$

$$z = \frac{\det\begin{bmatrix} 2 & 1 & 3 \\ 3 & 0 & 2 \\ 4 & 5 & 1 \end{bmatrix}}{15} = \frac{30}{2} = 2$$

7. In reality, the matrix A represents a function $A = A(t)$, because each value of t yields a different matrix. The expression $\det(A(t))$ is also a function of t. We know that $A(t)$ is invertible if and only if $\det(A(t))$ is nonzero, so we solve the equation

$$\det(A(t)) = 0$$

The computation is simplified by adding -2 times the second column to the first and -1 times the second column to the third. The resulting function of t is

$$2t^2 - 4t - 1 = 0.$$

Solving this equation using the quadratic formula, we obtain

$$t = (2 \pm \sqrt{6})/2.$$

8. We have

$$\det(A^2) = \det(A)\det(A) = (\det(A))^2 = 7,$$

so $\det(A) = \pm\sqrt{7}$. Since det(A) is nonzero, A is invertible.

9. We transpose the given matrix A, obtaining the matrix that follows on the left and then replace each entry by its cofactor, $(-1)^{i+j}\det(A_{ij})$, obtaining the adjoint of A on the right:

$$\begin{bmatrix} 3 & 1 & 2 \\ 2 & 1 & 5 \\ 1 & 2 & 1 \end{bmatrix} \rightarrow \begin{bmatrix} 2 & -1 & 0 \\ 1 & 7 & -5 \\ -3 & -11 & 10 \end{bmatrix}$$

10. The proof rests on the fact that the determinant of a product is the product of the determinants:

$$\begin{aligned} \det(B) &= \det(P^{-1}AP) \\ &= \det(P^{-1})\det(A)\det(P) \\ &= \det(A)\det(P^{-1})\det(P) \\ &= \det(A)\det(P^{-1}P) \\ &= \det(A)\det(I) \\ &= \det(A)(1) \\ &= \det(A) \end{aligned}$$

Solutions to Odd-Numbered Review Exercises

1. We can find this determinant fairly easily by adding the negative of the first row to the second to obtain

$$\det\begin{bmatrix}1 & 1 & 0\\ 1 & 0 & 1\\ 0 & 1 & 1\end{bmatrix} = \det\begin{bmatrix}1 & 1 & 0\\ 0 & -1 & 1\\ 0 & 1 & 1\end{bmatrix} = (1)\det\begin{bmatrix}-1 & 1\\ 1 & 1\end{bmatrix}$$

$$= (1)[(-1)(1) - (1)(1)] = -2$$

3. The given matrix is lower triangular, so the determinant is simply the product of its main diagonal entries: $(1)(2)(3) = 6$.

5. We first subtract the first row from each of the other two and then expand down the first column to obtain

$$\det\begin{bmatrix}1 & x & y\\ 1 & x^2 & y^2\\ 1 & x^3 & y^3\end{bmatrix} = \det\begin{bmatrix}1 & x & y\\ 0 & x^2-x & y^2-y\\ 0 & x^3-x & y^3-y\end{bmatrix} = \det\begin{bmatrix}x^2-x & y^2-y\\ x^3-x & y^3-y\end{bmatrix}$$

Factoring $x(x-1)$ out of the first column and $y(y-1)$ out of the second, we can simplify further so that the determinant becomes:

$$xy(x-1)(y-1)\det\begin{bmatrix}1 & 1\\ x+1 & y+1\end{bmatrix}$$

Finally, evaluating this 2×2 determinant above, we obtain the following expression for the original determinant: $xy(x-1)(y-1)(y-x)$

7. Again, there is no "best" strategy, but adding the last row to the first and expanding along the first row gives:

$$\det\begin{bmatrix}1 & 0 & 2 & 1\\ 0 & 1 & -1 & 0\\ 2 & 0 & 5 & 2\\ -1 & 0 & -2 & 0\end{bmatrix} = \det\begin{bmatrix}0 & 0 & 0 & 1\\ 0 & 1 & -1 & 0\\ 2 & 0 & 5 & 2\\ -1 & 0 & -2 & 0\end{bmatrix} = -(1)\det\begin{bmatrix}0 & 1 & -1\\ 2 & 0 & 5\\ -1 & 0 & -2\end{bmatrix}$$

Now, expanding down the second column, the determinant is:

$$-(1)(-1)\det\begin{bmatrix}2 & 5\\ -1 & -2\end{bmatrix} = 1$$

9. An inverse for a square matrix exists if and only if its determinant is nonzero. Thus, the invertible matrices are those of Exercises 1, 2, and 3, but not 4.

11. Since

$$1 = \det(I) = \det(AA^{-1}) = \det(A)\det(A^{-1}) = 2\det(A^{-1}),$$

it follows that $\det(A^{-1}) = 1/2$. Since

$$\det(A) = \det(A^T),$$

we also have $\det(A^T) = 2$. Moreover, an argument similar to the one just given implies that

$$\det(A^T)^{-1} = 1/2.$$

13. We first transpose the matrix and then replace each i, j^{th} entry by the cofactor of the i, j^{th} position: $(-1)^{i+j}\det(A_{ij})$

$$\begin{bmatrix} 1 & 1 & 0 \\ 1 & 0 & 1 \\ 0 & 1 & 1 \end{bmatrix} \rightarrow \begin{bmatrix} -1 & -1 & 1 \\ -1 & 1 & -1 \\ 1 & -1 & -1 \end{bmatrix} = \text{Adj}(A)$$

To find A^{-1}, multiply Adj(A) by the reciprocal of the determinant of the original matrix, $\det(A) = -2$; that is, $A^{-1} = (-1/2)\text{Adj}(A)$.

15. Proceeding as in Exercise 13:

$$\begin{bmatrix} 1 & 1 & 1 \\ 0 & 2 & 2 \\ 0 & 0 & 3 \end{bmatrix} \rightarrow \begin{bmatrix} 6 & 0 & 0 \\ -3 & 3 & 0 \\ 0 & -2 & 2 \end{bmatrix} = \text{Adj}(A)$$

Here, $\det(A) = 6$, so $A^{-1} = (1/6)\text{Adj}(A)$.

17. The determinant of the coefficient matrix is 3, so:

$$x = \frac{\det\begin{bmatrix} 1 & -2 \\ 1 & -1 \end{bmatrix}}{3} = \frac{1}{3}, \quad y = \frac{\det\begin{bmatrix} 1 & 1 \\ 2 & 1 \end{bmatrix}}{3} = -\frac{1}{3}$$

19. The determinant of the coefficient matrix is -2, so:

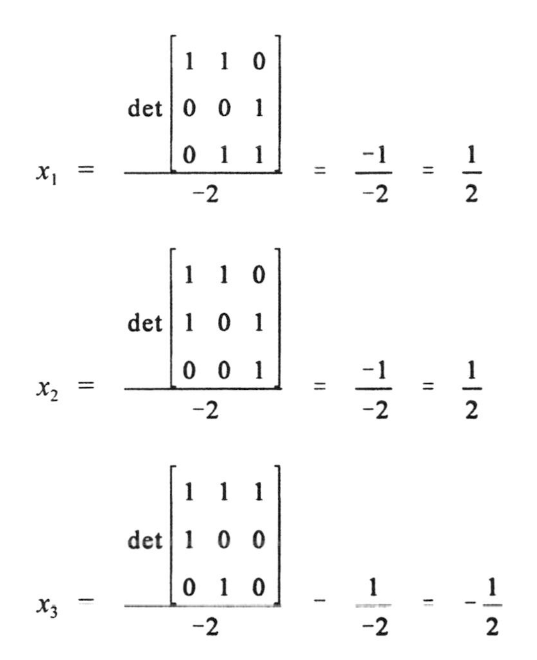

$$x_1 = \frac{\det\begin{bmatrix}1 & 1 & 0\\ 0 & 0 & 1\\ 0 & 1 & 1\end{bmatrix}}{-2} = \frac{-1}{-2} = \frac{1}{2}$$

$$x_2 = \frac{\det\begin{bmatrix}1 & 1 & 0\\ 1 & 0 & 1\\ 0 & 0 & 1\end{bmatrix}}{-2} = \frac{-1}{-2} = \frac{1}{2}$$

$$x_3 = \frac{\det\begin{bmatrix}1 & 1 & 1\\ 1 & 0 & 0\\ 0 & 1 & 0\end{bmatrix}}{-2} = \frac{1}{-2} = -\frac{1}{2}$$

21. The determinant of the coefficient matrix: is 0, so Cramer's rule does not apply.

Answers to Even-Numbered Exercises

SECTION 4.1

2. 2,2-minor is -8; 2,3-minor is -6; 2,2-cofactor is -8; 2,3-cofactor is 6.

4. -5

6. 0

8. 10

10. 5

12. 0

14. $k(ad - bc)$

16. $\lambda^2 - 3\lambda - 10$

18. 24

20. -15

24. $\det A = (\pm)(a_{1n})(a_{2,n-1})(a_{3,n-2})\ldots(a_{n1})$,

where the "+" sign is used if $n = 4k$ or $n = 4k + 1$ and "-", otherwise.

28. (0, 2, 3)

34. -6

SECTION 4.2

2. 6

4. 0

6. 0.001134

8. $(t - 3)(t^2 - t - 8)$

10. Matrices 1, 2, 3, 5, and 6 are invertible.

14. The solution is not unique.

16. The solution is unique.

20. $\det A' = (-1)^m \det A$

22. $(u - s)(u - t)(t - s)$

SECTION 4.3

2. Cannot be used; det A = 0.

4. $x = 5,\ y = -1,\ z = -5$

6. $x = -1,\ y = 1/2,\ z = -3$

8. Cannot be used; det A = 0.

10. $\text{Adj}(A) = \begin{bmatrix} -2 & 1 \\ -1 & 1 \end{bmatrix}, \quad A^{-1} = \begin{bmatrix} 2 & -1 \\ 1 & -1 \end{bmatrix}$

12. $\text{Adj}(A) = \begin{bmatrix} 10 & -4 & 2 \\ 2 & 8 & -4 \\ -7 & 5 & 3 \end{bmatrix}, \quad A^{-1} = \frac{1}{22}\begin{bmatrix} 10 & -4 & 2 \\ 2 & 8 & -4 \\ -7 & 5 & 3 \end{bmatrix}$

14. $\text{Adj}(A) = \begin{bmatrix} -3 & 2 & 4 \\ 2 & 1 & 2 \\ 4 & 2 & -3 \end{bmatrix}, \quad A^{-1} = \frac{1}{7}\begin{bmatrix} -3 & 2 & 4 \\ 2 & 1 & 2 \\ 4 & 2 & -3 \end{bmatrix}$

REVIEW EXERCISES

2. 1

4. 0

6. $2xy(x - y)$

8. 3

10. 1

12. Not possible. Both can be used if and only if A is square and invertible.

14. $\text{Adj}(A) = \begin{bmatrix} 1 & 0 & 1 \\ -1 & -1 & 0 \\ -1 & -1 & -1 \end{bmatrix} = A^{-1}$

16. $\text{Adj}(A) = \begin{bmatrix} 8 & 0 & -4 \\ -16 & 0 & 8 \\ 8 & 0 & -4 \end{bmatrix}$; A^{-1} does not exist.

18. $x = 1,\ y = 1$

20. Cramer's rule cannot be used.

22. Cramer's rule cannot be used.

CHAPTER 5

Independence and Basis in $\mathbf{R}^m$

Chapter Overview

The topics covered in this chapter are:

1. Definitions of *linear combination*, *linear independence/dependence*, *elementary vector* ($\mathbf{e}_i$, the m-vector with a 1 for its ith component, 0 everywhere else) in $\mathbf{R}^m$. [Section 5.1]
2. Computational procedures (involving row-reduction and determinants) for determining whether or not a set of vectors in $\mathbf{R}^m$ is linearly dependent and, if so, for determining the coefficients of dependence. [5.1]
3. The fact that a set of n vectors in $\mathbf{R}^m$ is linearly dependent if $n > m$. [5.1]
4. *Subspaces* of $\mathbf{R}^m$, including the *solution space* of a homogeneous system of linear equations and the *subspace generated by* a given set of vectors in $\mathbf{R}^m$. [5.2]
5. *Spanning sets* for a subspace of $\mathbf{R}^m$. [5.2]
6. *Basis* for a subspace of $\mathbf{R}^m$, including the standard basis for $\mathbf{R}^m$ and finding a basis for the solution space of a homogeneous linear system. [5.3]
7. Theorems concerning bases in $\mathbf{R}^m$: [5.3]
   - Every spanning set can be reduced to a basis.
   - Every linearly independent set can be extended to a basis.
   - Every basis contains the same number of vectors.

8. *Dimension* of a subspace of $\mathbf{R}^m$. [5.3]

9. The facts that in a subspace of $\mathbf{R}^m$ of dimension n, a set consisting of k vectors: [5.3]

   - must be linearly dependent if $k > n$;
   - cannot span the subspace if $k < n$.

10. The *row space* and *column space* of a matrix (whose dimensions are, respectively, the *row rank* and *column rank* of the matrix). [5.4]

11. The fact that the rank of a matrix (defined as the number of nonzero rows in its row-reduced echelon form), its row rank, and its column rank are always equal. [5.4]

Remarks

The main purpose of Chapter 5 is to provide a transition from the heavily computational material in the first four chapters to the more theoretical concepts that will appear in Chapter 6, *Vector Spaces*. Most students find the material dealing with vector spaces (the definition, linear independence, basis, and so on) to be very difficult, even overwhelming. To ameliorate this situation, Chapter 5 introduces the notions of subspace, linear independence, span, basis, and dimension in the friendly confines of $\mathbf{R}^m$.

You should be aware of two slight "discrepancies" between the definitions given in Chapters 5 and 6:

1. In Chapter 5, to enhance the geometric nature of the subject, a set is defined to be linearly dependent if one of its vectors can be expressed as a linear combination of the others; it is a theorem that a set is linearly dependent if and only if the appropriate vector equation has a nontrivial solution. In Chapter 6, however, we take the opposite (more conventional) approach: what was the theorem is now the definition and vice-versa.

2. A similar situation holds with regard to the two characterizations of *subspace*: in Chapter 5, a subspace of $\mathbf{R}^m$ is defined as a nonempty subset that is closed under addition and scalar multiplication. In Chapter 6, it is defined as a subset that is a vector space in its own right; the former definition becomes a theorem here.

Chapter 5 contains quite a bit of material, most of which will be discussed in a more general setting in Chapter 6. So, if you are pressed for time, you can easily skip certain topics and still provide some transition to abstract vector spaces. In particular, the following topics can be omitted without causing undue problems:

- The proofs of Theorems 4 and 5 of Section 5.1.
- Reducing a spanning set to a basis (in Section 5.3).
- Extending a linearly independent set to a basis (Section 5.3).
- Section 5.4, *Rank of a Matrix.*

We do suggest that you discuss the process of finding a basis for the solution space of a homogeneous linear system (Section 5.3). We will use this procedure in two important applications later in the text: determining the kernel of a linear transformation (Section 7.3) and finding the eigenspaces of a matrix (Section 8.1).

Most instructors find that the material in Chapter 5 (and ensuing chapters) cannot be covered as quickly as that of the previous chapters. So, be prepared to spend more time explaining new concepts and answering questions about them. This is a crucial juncture in the course, and those students who stumble here may never regain their equilibrium.

Solutions to Self-Test

1. We form the matrix with the given vectors as its columns and transform it to row-reduced echelon form:

$$\begin{bmatrix} 1 & 0 & 2 \\ 0 & 1 & 1 \\ 1 & -1 & 1 \\ 0 & 1 & 1 \end{bmatrix} \rightarrow \begin{bmatrix} 1 & 0 & 2 \\ 0 & 1 & 1 \\ 0 & 0 & 0 \\ 0 & 0 & 0 \end{bmatrix}$$

Since the third column of the row-reduced matrix is not an elementary one, by Theorem 4 of Section 5.1 the given set of vectors is linearly dependent. Inspecting the third column and applying Theorem 5 of Section 5.1, we see that

$$(2, 1, 1, 1) = (2)(1, 0, 1, 0) + (1)(0, 1, -1, 1)$$

2. The given set of vectors spans $\mathbf{R}^3$ if and only if an arbitrary vector, (a, b, c), can be written as a linear combination of the given vectors. This will be true if and only if there exist constants c_1, c_2, and c_3 such that

$$(a, b, c) = c_1(1, 1, 1) + c_2(0, 1, 2) + c_3(1, 2, 3)$$

Performing the scalar multiplications on the right side and equating like coefficients, we obtain the linear system:

$$\begin{aligned} c_1 \quad\quad\;\; + \;\; c_3 &= a \\ c_1 + \;\; c_2 + 2c_3 &= b \\ c_1 + 2c_2 + 3c_3 &= c \end{aligned}$$

By using Gauss-Jordan elimination, we see that this system fails to have a solution if $a = 0$, $b = 0$, and $c = 1$ (and for many other choices of a, b, and c as well). Thus, the given set of vectors does not span $\mathbf{R}^3$.

Note: An easier way to solve this problem is to invoke the methods of Section 5.3. For example, letting A be the matrix whose columns are the given vectors, we see that $\det A = 0$. But then, by Theorem 6 of Section 5.3, the given set is not a basis for $\mathbf{R}^3$, and therefore (by Theorem 5iii), does not span $\mathbf{R}^3$.

3. The given set, call it S, is certainly a nonempty subset of $\mathbf{R}^4$. To show that it is a subspace, we must verify the two conditions of the definition. Let $\mathbf{u} = (a_1 + b_1, a_1 - b_1, c_1, 0)$ and $\mathbf{v} = (a_2 + b_2, a_2 - b_2, c_2, 0)$ be arbitrary vectors in S and let k be a constant. Then:

i. $$\begin{aligned} \mathbf{u} + \mathbf{v} &= (a_1 + b_1, a_1 - b_1, c_1, 0) \;+\; (a_2 + b_2, a_2 - b_2, c_2, 0) \\ &= (a_1 + b_1 + a_2 + b_2, a_1 - b_1 + a_2 - b_2, c_1 + c_2, 0) \\ &= ((a_1 + a_2) + (b_1 + b_2), (a_1 + a_2) - (b_1 + b_2), c_1 + c_2, 0) \end{aligned}$$

which is (of the form of) a vector in S.

ii. $$\begin{aligned} k\mathbf{u} &= k(a_1 + b_1, a_1 - b_1, c_1, 0) \\ &= (k(a_1 + b_1), k(a_1 - b_1), kc_1, 0) \\ &= (ka_1 + kb_1, ka_1 - kb_1, kc_1, 0) \end{aligned}$$

which is (of the form of) a vector in S. Thus, whenever $\mathbf{u}$ and $\mathbf{v}$ are in S, so are $\mathbf{u} + \mathbf{v}$ and $k\mathbf{u}$ (for arbitrary scalar k). By definition, S is a subspace of $\mathbf{R}^4$.

4. The set of all vectors with first component equal to 1 is not a subspace of $\mathbf{R}^m$; it does not contain the zero vector. (You can also show that the given set is not a subspace by adding two vectors in it or by multiplying a vector in it by a scalar other than 1. In both cases, the resulting vector will not be in the given set.)

5. Each of the vectors in the given set, call it T, is in the subspace **S** of Problem 3. We must show that T is linearly independent and that it spans **S**. To show the former, construct the matrix whose columns are the vectors in T and transform it to row-reduced echelon form:

$$\begin{bmatrix} 1 & 1 & 0 \\ 1 & -1 & 0 \\ 0 & -0 & 1 \\ 0 & 0 & 0 \end{bmatrix} \rightarrow \begin{bmatrix} 1 & 0 & 0 \\ 0 & 1 & 0 \\ 0 & 0 & 1 \\ 0 & 0 & 0 \end{bmatrix}$$

Since the row-reduced matrix consists solely of distinct elementary columns, T is linearly independent.

To prove that T spans **S**, just take an arbitrary vector in **S** and show that it can be written as a linear combination of the vectors of T:

$$(a + b, a - b, c, 0) = a(1, 1, 0, 0) + b(1, -1, 0, 0) + c(0, 0, 1, 0)$$

6. Since a basis for the subspace of Problem 3 (as demonstrated in Problem 5) contains 3 vectors, the dimension of this subspace is 3.

7. Notice that the matrix A is in row-reduced echelon form. The corresponding system of equations is:

$$\begin{aligned} x \quad\quad + z + 2w &= 0 \\ y + z + \ \ w &= 0 \end{aligned}$$

Letting $z = s$ and $w = t$, we obtain the two-parameter family of solutions: $(-s - 2t, -s - t, s, t)$. Now, setting s and t equal to the following values yields (by Theorem 4 of Section 5.3) the indicated basis vectors:

$s = 1,\ t = 0$: $(-1, -1, 1, 0)$
$s = 0,\ t = 1$: $(-2, -1, 0, 1)$

8. We can determine the truth or falsity of these statements by applying Theorem 5 of Section 5.3.

a. True; there are too few vectors in T (2) to span a 3-dimensional space.

b. False; if these vectors are linearly dependent, they will not be a basis for S.

c. True; there are too many vectors (4) in T for it to be a linearly independent set in a 3-dimensional space.

9. a. To find a basis for the column space of B, we transform this matrix to row-reduced echelon form:

$$\begin{bmatrix} 1 & 0 & -1 & -2 \\ 0 & 1 & 2 & 3 \\ 0 & 0 & 0 & 0 \end{bmatrix}$$

The two columns that were transformed into the elementary ones form a basis for the column space: $\{(1, 0, 1), (2, 1, 1)\}$. We see that its dimension is 2.

b. To find a basis for the row space of B, we can transform B^T to row-reduced echelon form:

$$\begin{bmatrix} 1 & 0 & 1 \\ 2 & 1 & 1 \\ 3 & 2 & 1 \\ 4 & 3 & 1 \end{bmatrix} \rightarrow \begin{bmatrix} 1 & 0 & 2 \\ 0 & 1 & -1 \\ 0 & 0 & 0 \\ 0 & 0 & 0 \end{bmatrix}$$

As you can see, the first two columns of B^T have been transformed to row-reduced echelon form. Thus, the first two rows of B form the 2-dimensional row space generated by $\{(1, 2, 3, 4), (0, 1, 2, 3)\}$.

10. We must show that T spans S and is linearly independent. Since T generates S, by definition it spans this subspace. Suppose T is not linearly independent. Then, there are scalars c_1 and c_2, not both zero, such that

$$c_1\mathbf{u} + c_2\mathbf{v} = \mathbf{0}.$$

But this implies that for this c_1 and c_2,

$$c_1\mathbf{u} + c_2\mathbf{v} + 0\mathbf{w} = \mathbf{0}.$$

Thus, we have a linear combination of **u**, **v**, and **w** equal to **0**, with at least one coefficient not equal to 0, which contradicts the linear independence of this set. Therefore, **u** and **v** form a linearly independent set.

Solutions to Odd-Numbered Review Exercises

2. We form the matrix A with the given vectors as columns:

$$\begin{bmatrix} 1 & 1 & 0 \\ 0 & 2 & 2 \\ 0 & 0 & 3 \end{bmatrix}$$

Since det A = 6 ≠ 0, by Theorem 3 of Section 5.1, the given vectors are linearly independent.

3. We form the matrix with the given vectors as columns and transform it to row-reduced echelon form:

$$\begin{bmatrix} 1 & -1 & 3 \\ 0 & 1 & -1 \\ 1 & 2 & 0 \\ 0 & -1 & 1 \end{bmatrix} \rightarrow \begin{bmatrix} 1 & 0 & 2 \\ 0 & 1 & -1 \\ 0 & 0 & 0 \\ 0 & 0 & 0 \end{bmatrix}$$

Since the third column of the latter matrix is not an elementary one, the given set is linearly dependent, and (from its last column):

$$(3, -1, 0, 1) = 2(1, 0, 1, 0) + (-1)(-1, 1, 2, -1)$$

5. We check the two conditions of the definition:

i. $(0, x_1, -2x_1, x_1 + 2y_1) + (0, x_2, -2x_2, x_2 + 2y_2)$
$= (0, x_1 + x_2, -2x_1 - 2x_2, x_1 + 2y_1 + x_2 + 2y_2)$
$= (0, x_1 + x_2, -2(x_1 + x_2), (x_1 + x_2) + 2(y_1 + y_2)),$

which is a vector in the given set.

ii. $c(0, x, -2x, x + 2y) = (0, cx, c(-2x), c(x + 2y))$
$= (0, cx, -2(cx), cx + 2(cy)),$

which is a vector in the given set. Thus, the given set of vectors satisfies the definition and is a subspace of $\mathbf{R}^4$.

7. We only need show that one condition of the definition is violated (even though, in this case, neither condition holds). Taking $\mathbf{x} = (0, 0, -1/2)$, we see that $\mathbf{x} + (0, 0, 1)$ is nonzero. Thus, $(0, 0, -1/2)$ is in the given set. However,

$$2\mathbf{x} + (0, 0, 1) = (0, 0, -1) + (0, 0, 1) = \mathbf{0},$$

so $2\mathbf{x}$ is not in the given set. Thus, condition *ii* is not satisfied and the given set is not a subspace of $\mathbf{R}^3$.

9. We first find the solution space, **S**, of $A\mathbf{x} = \mathbf{0}$. By solving this system, we see that an arbitrary vector in **S** has the form $(-t, t, t)$, where t is a scalar. Now,

$$(-t, t, t) = t(-1, 1, 1),$$

so $\mathbf{x} = (-1, 1, 1)$ spans **S**.

11. Notice that these are the vectors of Exercise 1. Since they are linearly independent, and there are three of them, they form a basis for $\mathbf{R}^3$.

13. Each of the given vectors is in the given subspace, **S**. Moreover, **S** is 2-dimensional (a basis consists of $(1, 1, 1, 0)$ and $(0, 1, -1, 0)$), and the two vectors in the given set are linearly independent (they are not scalar multiples of one another). Thus, the given set is a basis for **S**.

15. A general vector in the given subspace (call it **S**) has the form $(x, x+y+z, y, z)$. Now, take:

$$\begin{aligned} &x = 1, y = 0, z = 0 \text{ to get } (1, 1, 0, 0); \\ &x = 0, y = 1, z = 0 \text{ to get } (0, 1, 1, 0); \\ &x = 0, y = 0, z = 1 \text{ to get } (0, 1, 0, 1). \end{aligned}$$

These vectors are linearly independent and span **S**. To see the latter, notice that

$$(x, x+y+z, y, z) = x(1, 1, 0, 0) + y(0, 1, 1, 0) + z(0, 1, 0, 1).$$

Thus, $\{(1, 1, 0, 0), (0, 1, 1, 0), (0, 1, 0, 1)\}$ is a basis for **S**, and $\dim \mathbf{S} = 3$.

17. Let $\mathbf{x} = (x, y, z)$. Then,

$$0 = \mathbf{x} \cdot (1, 2, 3) = x + 2y + 3z.$$

or $x = -2y - 3z$. Thus, a general vector in the given subspace (call it **S**) has the form $(-2y - 3z, y, z)$. Proceeding as in the solution to Exercise 15, we take:

$$\begin{aligned} &y = 1, z = 0 \text{ to get } (-2, 1, 0); \\ &y = 0, z = 1 \text{ to get } (-3, 0, 1). \end{aligned}$$

Thus, a basis for **S** is $\{(-2, 1, 0), (-3, 0, 1)\}$ and the dimension of **S** is 2.

19. The row space of B is generated by the set of vectors $\{(1, 1, 1, 1, 1), (0, 1, 1, 1, 1), (0, 0, 1, 1, 1)\}$. Since this set is linearly independent, it is a basis for the row space, and the dimension of this space is 3.

21. We reduce the given set to a basis by transforming the matrix on the left to its row-reduced echelon form on the right:

$$\begin{bmatrix} 1 & -1 & 2 & 2 \\ -1 & -1 & -1 & -3 \\ 2 & 4 & 1 & 7 \\ 0 & -2 & 1 & -1 \end{bmatrix} \rightarrow \begin{bmatrix} 1 & 0 & 3/2 & 5/2 \\ 0 & 1 & -1/2 & 1/2 \\ 0 & 0 & 0 & 0 \\ 0 & 0 & 0 & 0 \end{bmatrix}$$

The two vectors, (1, -1, 2, 0) and (-1, -1, 4, -2), that were transformed into elementary ones form a basis for the subspace, **S**, generated by the given vectors and the dimension of **S** is 2.

Answers to Even-Numbered Exercises

SECTION 5.1

4. $x_1 + x_2 + x_3 = 0$

6. Independent

8. (0, 0, 0, 0) = 0(-1, -2, 2, 1) + 0(1, 2, 3, 4)

10. Independent

12. (4, -4, 1) = 1(2, 0, 1) + 2(1, -2, 0)

14. (0, -1, -2, -3) = (-1)(0, 1, 2, 3)

16. $\mathbf{v}_4 = (-1)\mathbf{v}_1 + 3\mathbf{v}_3$

18. Independent

20. Dependent by Theorem 2

22. Dependent by the definition (the third vector is the sum of the first two)

24. Independent by Theorem 3

26. Dependent by the definition (the third vector is a scalar multiple of the first)

SECTION 5.2

14. Spans

16. Spans

20. $\{\mathbf{e}_1, \mathbf{e}_2, ..., \mathbf{e}_m\}$

22. $\{\mathbf{0}\}$

24. {(-1, 1, 0), (-1, 0, 1)}

SECTION 5.3

2. Is a basis

4. Not a basis (not even a subset)

6. Not a basis

8. Is a basis

10. {(1, 0, 1, 0), (0, -1, 1, 2), (1, 1, 1, 1)}

12. {(1, -1, 0), (0, 1, 1), (1, 0, 2)}

14. {(-1, 1, 0, 0), (1, -1, 1, 0), (0, 0, 1, 2), (1, 0, 0, 0)}

16. {(-1, -1, 2, 1), (2, 1, -1, -2), (0, -1, -4, 0), (1, 0, 0, 0)}

18. Two vectors cannot span a 3-dimensional subspace.

20. By Theorem 5iii, with $m = 4$, $n = 2$

22. {(-1, 0, 1, 0, 0), (3, 1, 0, 1, 0), (-2, -1, 0, 0, 1)}; 3

24. {(1, 0, 1), (1, 1, 0)}; 2

26. {(0, 1, 0, 1), (0, 0, 1, -1)}; 2

28. {(1, -1, 0), (0, 1, 1)}; 2

30. 2

32. **a.** 3 **b.** 1

34. **a.** 2 **b.** 2

SECTION 5.4

8. {(1, 2), (2, 1)}

10. {0, -1, 2)}

12. {(1, 1, 1), (-1, 1, 0), (1, 0, 0)}

14. {(1, 2, 1), (2, 1, -1)}

16. {(-1, 2)}

18. {(1, -1, 1, 0), (1, 1, 0, 0), (1, 0, 0, 1)}

24. {(1, 2, 0), (0, 0, 1)}

26. {(1, 0, -1)}

28. {(1, -1, 3), (0, 1, 1), (1, 1, 0)}

REVIEW EXERCISES

2. (2, 1, 1) = 2(1, 2, -1) + (-3)(0, 1, -1)

4. Independent

6. Not a subspace

8. Is a subspace

10. Does not span

12. Is a basis

14. Not a basis

16. {(0, 0, 1, 0, 1), (0, 0, 0, 1, 1)}; 2

18. {(1, -2, 1, 0), (2, -3, 0, 1)}; 2

20. {(1, 0, 0), (1, 1, 0), (1, 1, 1)}; 3

22. {(1, 2, 3), (3, 2, 1), (1, 0, 0)}

CHAPTER 6

Vector Spaces

Chapter Overview

The topics covered in this chapter are:

1. Definition of *vector space* (over the real numbers) and *subspace*; examples, including: [Section 6.1]

   - $\mathbf{R}^m$: Euclidean m-space under the operations of vector addition and multiplication by a scalar.
   - $\mathbf{M}^{m,n}$: The set of all $m \times n$ matrices under matrix addition and multiplication by a scalar.
   - **P**: The set of all polynomials under the operations of polynomial addition and multiplication by a real number.
   - $C(a, b)$: The set of all real-valued functions continuous on an interval (a, b) under the operations of function addition and multiplication by a real number.
   - $\mathbf{P}_n$: The set of all polynomials of degree less than or equal to n under the operations of polynomial addition and multiplication by a real number.

2. For an arbitrary vector space: [6.2]

   - *Linear dependence* and *independence* of a set of vectors.
   - *Span* of a set of vectors; the subspace *generated by* a set of vectors.
   - *Basis* for a vector space (including the standard bases for the spaces listed

in item 1, except **C**(a, b)); basic theorems about bases:

a. Every spanning set can be reduced to a basis.
b. Every linearly independent set can be extended to a basis.
c. All bases for a given vector space contain the same number of vectors.

- *Dimension* of a vector space (including the dimensions of the spaces listed in item 1); basic theorems about k-element subsets, S, of a vector space **V** of dimension n:

  a. If $k > n$, then S is linearly dependent;
  b. If $k < n$, then S cannot span **V**;
  c. If $k = n$, then S is linearly independent if and only if it spans **V**.

3. *Coordinate vectors*; using coordinate vectors to translate problems concerning abstract vector spaces into the context of $\mathbf{R}^m$. [6.3]
4. *Inner products* for an abstract vector space; properties (including *norms* and the Cauchy-Schwarz inequality). [6.4]
5. Orthogonal and orthonormal sets and bases. [6.4]
6. The Gram-Schmidt process for constructing an orthonormal basis. [6.4]
7. Application of inner product spaces to approximating continuous functions and Fourier series. [6.5]

Remarks

The topic of vector spaces (including linear independence, basis, and dimension) is arguably the most difficult topic in a beginning linear algebra course. (Linear transformations, which are covered in Chapter 7, are no picnic either.) Doing numerous examples should help ease the way here. It is also important to show the students how this new material is just an extension of the now familiar concepts of Chapter 5.

In addition to the definition of vector space and subspace, Section 6.1 supplies several important examples of vector spaces, described in item 1 of the Chapter Overview above. The vector spaces $\mathbf{R}^m$, $\mathbf{M}^{m,n}$, and $\mathbf{P}_n$ will appear throughout the rest of the text. Because **P** (the space of *all* polynomials) is infinite-dimensional, it is less useful in an elementary linear algebra course.

The vector space **C**(a, b), together with its extensions to functions that are continuous on $(-\infty,\infty)$, differentiable on (a, b), and so on, is used in two application sections: Section 6.5 (approximation of continuous functions) and Section 7.4 (differential equations). The introduction of $\mathbf{C}(a, b)$ and its extensions is completely optional; this material can be omitted without causing any future problems.

Sections 6.1 and 6.2 form the heart of this chapter and are crucial for an understanding of Chapter 7. Sections 6.3 and 6.4, on the other hand, contain less important material and can be skipped if time is a problem. Section 6.3, *Coordinate Vectors*, demonstrates that finite-dimensional vector spaces are very closely related to $\mathbf{R}^m$ (although the topic of isomorphism is not discussed). This material is used only in Sections 7.5 and 7.6, which discuss the matrix of a general linear transformation and the change of basis process. Section 6.4 contains several subtopics (see the Chapter Overview above), all related to the idea of an inner product on a vector space. This material appears again only in Sections 6.5 and 8.5, both of which are optional.

Solutions to Self-Test

1. The given set is not a vector space because it is not closed under scalar multiplication (property (b) of the definition). For a specific example, take $(a, b) = (1, 1)$ and $c = -1$. Then, (a, b) is in the given set but $c(a, b) = (-1,-1)$ is not.

2. The given set, call it S, is a vector space. Since S is a nonempty subset of the vector space of all 2×2 matrices, by Theorem 2 of Section 6.1 all we need show is that properties (a) and (b) of the definition are satisfied.

   (a) Let

$$A = \begin{bmatrix} u & 0 \\ v & w \end{bmatrix} \quad \text{and} \quad B = \begin{bmatrix} x & 0 \\ y & z \end{bmatrix}$$

   be arbitrary elements of S. Then,

$$A + B = \begin{bmatrix} u + x & 0 \\ v + y & w + z \end{bmatrix}$$

   is also in S (it has a 0 in its upper-right corner). Thus, S is closed under addition — property (a) is satisfied.

(b) Let

$$A = \begin{bmatrix} u & 0 \\ v & w \end{bmatrix}$$

be an arbitrary element in S. Then

$$cA = \begin{bmatrix} cu & 0 \\ cv & cw \end{bmatrix}$$

is also an element of S. Thus, S is closed under scalar multiplication — property (b) is satisfied — and S is a vector space.

3. The given set, call it S, is a subset of **P**, the vector space of all polynomials. Then, S is a vector space (by Theorem 2 of Section 6.1) because:

(a) S is closed under addition since adding two polynomials with missing x^2 terms results in a polynomial with no x^2 term.

(b) S is closed under scalar multiplication since multiplying a polynomial with no x^2 term by a scalar results in a polynomial with no x^2 term.

4. This set is not a vector space; it is not closed under addition (property (a) fails to hold). To see this, take

$$\mathbf{p}(x) = x^4 + x + 10 \quad \text{and} \quad \mathbf{q}(x) = -x^4.$$

Then, $(\mathbf{p}+\mathbf{q})(x) = x + 10$, which does not have degree equal to 4 and therefore is not in the given set.

5. We have to find scalars c_1, c_2, and c_3 such that

$$1 + x - x^2 = c_1(1 + x + x^2) + c_2(x - x^2) + c_3(1 - x^2)$$

Performing the scalar multiplications on the right side and equating like powers of x results in the linear system:

$$\begin{aligned} c_1 \qquad + c_3 &= 1 \\ c_1 + c_2 \qquad &= 1 \\ c_1 - c_2 - c_3 &= -1 \end{aligned}$$

This system has the solution $c_1 = 1/3$, $c_2 = 2/3$, $c_3 = 2/3$. Thus:

$$1 + x - x^2 = (1/3)(1 + x + x^2) + (2/3)(x - x^2) + (2/3)(1 - x^2)$$

6. The four polynomials given in Exercise 5 are all elements of the vector space $\mathbf{P}_2$, of all polynomials of degree less than or equal to 2. Thus, we have 4 vectors in a vector space of dimension 3, and by Theorem 7 of Section 6.2, they must be linearly dependent.

7. Recall that the standard basis for $\mathbf{M}^{2,2}$ consists of the matrices:

$$\begin{bmatrix} 1 & 0 \\ 0 & 0 \end{bmatrix}, \quad \begin{bmatrix} 0 & 1 \\ 0 & 0 \end{bmatrix}, \quad \begin{bmatrix} 0 & 0 \\ 1 & 0 \end{bmatrix}, \quad \begin{bmatrix} 0 & 0 \\ 0 & 1 \end{bmatrix}$$

Relative to this standard basis, the given matrices have the following coordinate vectors:

$$A = \begin{bmatrix} 1 & 1 \\ 1 & 1 \end{bmatrix} \rightarrow (1, 1, 1, 1)$$

$$A_1 = \begin{bmatrix} 0 & 1 \\ 1 & 1 \end{bmatrix} \rightarrow (0, 1, 1, 1)$$

$$A_2 = \begin{bmatrix} 1 & 0 \\ 1 & 1 \end{bmatrix} \rightarrow (1, 0, 1, 1)$$

$$A_3 = \begin{bmatrix} 1 & 1 \\ 0 & 1 \end{bmatrix} \rightarrow (1, 1, 0, 1)$$

$$A_4 = \begin{bmatrix} 1 & 1 \\ 1 & 0 \end{bmatrix} \rightarrow (1, 1, 1, 0)$$

Now, to express the matrix A as a linear combination of the other four matrices, we just express its coordinate vector as a linear combination of the other four coordinate vectors. To do this, we transform the following matrix (on the left) to its row-reduced echelon form on the right:

$$\begin{bmatrix} 0 & 1 & 1 & 1 & 1 \\ 1 & 0 & 1 & 1 & 1 \\ 1 & 1 & 0 & 1 & 1 \\ 1 & 1 & 1 & 0 & 1 \end{bmatrix} \rightarrow \begin{bmatrix} 1 & 0 & 0 & 0 & 1/3 \\ 0 & 1 & 0 & 0 & 1/3 \\ 0 & 0 & 1 & 0 & 1/3 \\ 0 & 0 & 0 & 1 & 1/3 \end{bmatrix}$$

From the last column in the row-reduced echelon form, we see that

$$A = (1/3)A_1 + (1/3)A_2 + (1/3)A_3 + (1/3)A_4$$

8. We must show that for all scalars a and b, $(a\mathbf{v}, b\mathbf{w}) = 0$. To do so, we use the rules for manipulating inner products given in Theorem 1a of Section 6.4:

$$(a\mathbf{v}, b\mathbf{w}) = a(\mathbf{v}, b\mathbf{w}) = ab(\mathbf{v}, \mathbf{w}) = ab(0) = 0$$

9. For the given functions, we have:

$$\mathbf{f}(0) = 0, \quad \mathbf{f}(1) = -1, \quad \mathbf{f}(2) = 0,$$

$$\mathbf{g}(0) = -1, \quad \mathbf{g}(1) = 0, \quad \mathbf{g}(2) = 1$$

Thus,

$$\mathbf{f}(0)\mathbf{g}(0) + \mathbf{f}(1)\mathbf{g}(1) + \mathbf{f}(2)\mathbf{g}(2) = (0)(-1) + (-1)(0) + (0)(1) = 0;$$

that is, $(\mathbf{f}, \mathbf{g}) = 0$, as desired.

10. We apply the Gram-Schmidt process to $\mathbf{v}_1$, $\mathbf{v}_2$, and $\mathbf{v}_3$. Since the first two of these vectors are orthogonal:

$$\begin{aligned} \mathbf{w} &= (1, 1, 1, 1) - \frac{(1, 1, 1, 1) \cdot (1, 1, 1, 0)}{(1, 1, 1, 0) \cdot (1, 1, 1, 0)}(1, 1, 1, 0) \\ &\quad - \frac{(1, 1, 1, 1) \cdot (1, 0, -1, 1)}{(1, 0, -1, 1) \cdot (1, 0, -1, 1)}(1, 0, -1, 1) \\ &= (1, 1, 1, 1) - (1)(1, 1, 1, 0) - (1/3)(1, 0, -1, 1) \\ &= (-1/3, 0, 1/3, 2/3) \end{aligned}$$

Solutions to Odd-Numbered Chapter Exercises

1. The quickest way to see that this set is not a vector space is to notice that it does not contain the zero polynomial, all of whose coefficients are zero. Thus, property (e) of the definition fails to hold. Since there is no **0** vector, there cannot be negatives, and property (f) fails as well.

Moreover, the set is not closed under addition (if we add two elements of the given set, the sum of the coefficients of the result is always 2) or multiplication by a scalar (for example, multiplying any element by 0 gives **0**). Hence, properties (a) and (b) also fail to hold.

3. This set (call it S) is a vector space under the given operations. By Theorem 2 of Section 6.1, we need only show that it is a subspace of **C**[0, 3]; that it is closed under addition and multiplication by a scalar. To do so, let **f** and **g** be vectors in S and let c be a scalar. Then:

i) For any x in [1, 2], $(\mathbf{f}+\mathbf{g})(x) = \mathbf{f}(x)+\mathbf{g}(x) = 0+0 = 0$. Thus, $\mathbf{f}+\mathbf{g}$ is in S.

ii) For any x in [1, 2], $(c\mathbf{f})(x) = c(\mathbf{f}(x)) = c(0) = 0$, so $c\mathbf{f}$ is in S.

5. This is not a vector space; the entry in the second row, second column causes the problem. We will show that the given set (call it S) is not closed under addition or multiplication by a scalar.

Taking $a=1$ and $b=2$, add this matrix to itself:

$$\begin{bmatrix} 1 & 2 & -1 \\ 5 & 2 & 1 \end{bmatrix} + \begin{bmatrix} 1 & 2 & -1 \\ 5 & 2 & 1 \end{bmatrix} = \begin{bmatrix} 2 & 4 & -2 \\ 10 & 4 & 2 \end{bmatrix}$$

Now, if the matrix sum were in S, from the first two entries in the first row, we must have $a=2$ and $b=4$. But that would mean that the entry in the second row, third column should be 8, which it is not. Thus, S is not closed under addition.

The argument above also shows that if a matrix A is in S, then 2A is not necessarily in S. Thus, S is not closed under multiplication by a scalar either. (All of the rest of the properties do hold.)

7. We will show that the given set is linearly independent in two ways: first, directly from the definition and then, by using coordinate vectors.

Direct Method: Let c_1, c_2, and c_3 be scalars and consider the equation:

$$c_1(x+2x^2+3x^3) + c_2(-x+2x^2-x^3) + c_3(x+x^3) = \mathbf{0}$$

Now, multiplying out by the scalars and combining like coefficients, we obtain

$$(c_1-c_2+c_3)x + (2c_1+2c_2)x^2 + (3c_1-c_2+c_3)x^3 = \mathbf{0}$$

Since the polynomial on the left is equal to the zero polynomial, each coefficient must be 0. This gives the linear system:

$$\begin{aligned} c_1 - c_2 + c_3 &= 0 \\ 2c_1 + 2c_2 \quad\quad &= 0 \\ 3c_1 - c_2 + c_3 &= 0 \end{aligned}$$

Since the only solution of this system is the trivial one, $\mathbf{c} = \mathbf{0}$, S is linearly independent.

Using coordinate vectors: The coordinate vectors for each polynomial in S (relative to the standard basis for $\mathbf{P}_3$) are (1, 2, 3), (-1, 2, -1), and (1, 0, 1). Now, the given set S is linearly independent if and only if this set of vectors in $\mathbf{R}^3$ is linearly independent. We can test the latter by calculating

$$\det \begin{bmatrix} 1 & -1 & 1 \\ 2 & 2 & 0 \\ 3 & -1 & 1 \end{bmatrix} = -4$$

Since this determinant is not zero, its columns (and hence S) are linearly independent.

9. This set is linearly dependent because the third vector is a scalar multiple of the first:

$$-2x + 2\cos x = (-2)(x - \cos x)$$

11. The given set is orthogonal if and only if the three pairs of distinct vectors in it are orthogonal. To check this, we find the dot products:

$$(2, -1, 1, 0) \cdot (0, 1, 1, 0) = 0$$
$$(2, -1, 1, 0) \cdot (1, 1, -1, -2) = 0$$
$$(0, 1, 1, 0) \cdot (1, 1, -1, -2) = 0$$

Thus, the given set is orthogonal.

13. Since there are only two vectors in the given set, we need only check one inner product:

$$\int_0^1 (1)(1 - 2x)\, dx = \left[x - x^2\right]_0^1 = 0$$

Thus, the given set is orthogonal.

15. Again, there are only two vectors in the given set, so we need only check one inner product. Letting

$$\mathbf{p}(x) = x - 2 \quad \text{and} \quad \mathbf{q}(x) = x^2 - x,$$

we have:

$$\begin{aligned}(\mathbf{p}, \mathbf{q}) &= \mathbf{p}(0)\mathbf{q}(0) + \mathbf{p}(1)\mathbf{q}(1) + \mathbf{p}(2)\mathbf{q}(2) \\ &= (-2)(0) + (-1)(0) + (0)(2) \\ &= 0\end{aligned}$$

Thus, the given set is orthogonal.

17. This is not an inner product; it violates the condition that $(\mathbf{p}, \mathbf{p}) = 0$ must imply that $\mathbf{p} = \mathbf{0}$. To understand why this is so, notice that $(\mathbf{p}, \mathbf{p}) = 0$ means that

$$[\mathbf{p}(1)]^2 + [\mathbf{p}(2)]^2 = 0,$$

which implies that $\mathbf{p}(1) = 0$ and $\mathbf{p}(2) = 0$. This only imposes two conditions on a polynomial in $\mathbf{P}_2$. Since the latter has three degrees of freedom, this is not enough to force $\mathbf{p}$ to be $\mathbf{0}$.

To show that this condition is violated, we construct a second degree polynomial which is 0 at $x = 1$ and $x = 2$, namely

$$\mathbf{p}(x) = (x - 1)(x - 2).$$

Then, $(\mathbf{p}, \mathbf{p}) = 0$, but $\mathbf{p}$ is not the zero polynomial.

19. This is not an inner product. It also violates the condition that $(A, A) = 0$ must imply that A is the zero matrix. Any matrix A whose diagonal entries are all 0, but which has at least one nonzero off-diagonal entry serves as a counter-example.

21. We first find an orthogonal basis, $\{\mathbf{v}_1, \mathbf{v}_2, \mathbf{v}_3\}$, for the subspace generated by the given set:

$$\mathbf{v}_1 = (1, 1, 1, 0)$$

$$\begin{aligned}\mathbf{v}_2 &= (1, 1, 0, 1) - \frac{(1, 1, 0, 1)\cdot(1, 1, 1, 0)}{(1, 1, 1, 0)\cdot(1, 1, 1, 0)}(1, 1, 1, 0) \\ &= (1, 1, 0, 1) - (2/3)(1, 1, 1, 0) = (1/3, 1/3, -2/3, 1)\end{aligned}$$

$$\begin{aligned}\mathbf{v}_3 &= (1, 0, 1, 1) - \frac{(1, 0, 1, 1)\cdot(1, 1, 1, 0)}{(1, 1, 1, 0)\cdot(1, 1, 1, 0)}(1, 1, 1, 0) \\ &\quad - \frac{(1, 0, 1, 1)\cdot(1/3, 1/3, -2/3, 1)}{(1/3, 1/3, -2/3, 1)\cdot(1/3, 1/3, -2/3, 1)}(1/3,1/3,-2/3,1) \\ &= (1, 0, 1, 1) - (2/3)(1, 1, 1, 0) - (2/5)(1/3, 1/3, -2/3, 1) \\ &= (3/15, -12/15, 9/15, 9/15)\end{aligned}$$

We now normalize each of the $\mathbf{v}_i$ (divide $\mathbf{v}_i$ by its length) to obtain the desired orthonormal basis:

$$\{ (1/\sqrt{3})(1, 1, 1, 0),\ (1/\sqrt{15})(1, 1, -2, 3),\ (1/\sqrt{35})(1, -4, 3, 3) \}$$

23. In words, this (alleged) inner product is computed by multiplying corresponding entries in the two matrices and their row and column numbers, and then summing the results. For simplicity of notation in the following argument, we will use

$$\sum ij\, a_{ij}\, b_{ij} \quad \text{to mean} \quad \sum_{i=1}^{m} \sum_{j=1}^{n} ij\, a_{ij}\, b_{ij}$$

We now verify the properties of the definition.

a. (A, B) = (B, A) holds because

$$\sum ij\, a_{ij}\, b_{ij} = \sum ij\, b_{ij}\, a_{ij}$$

due to the commutativity of the real numbers under multiplication.

b.
$$\begin{aligned} (A, B + C) &= \sum ij\, a_{ij} (b_{ij} + c_{ij}) \\ &= \sum [ij\, a_{ij}\, b_{ij} + ij\, a_{ij}\, c_{ij}] \\ &= \sum ij\, a_{ij}\, b_{ij} + \sum ij\, a_{ij}\, c_{ij} \\ &= (A, B) + (A, C) \end{aligned}$$

by just rearranging terms in the sum.

c.
$$\begin{aligned} (cA, B) &= \sum ij\, c\, a_{ij}\, b_{ij} \\ &= c\, [\sum ij\, a_{ij}\, b_{ij}] \\ &= c\, (A, B) \end{aligned}$$

d. Each term in

$$(A, A) = \sum ij\, a_{ij}\, a_{ij} = \sum ij\, (a_{ij})^2$$

is nonnegative (i and j are positive and $(a_{ij})^2$ is greater than or equal to zero), so their sum is nonnegative as well. Moreover, if (A, B) = 0, then each term must be zero. But i and j are strictly positive, so $(a_{ij})^2 = 0$, which implies that $a_{ij} = 0$, as desired.

Answers to Even-Numbered Exercises

SECTION 6.1

2. Not a vector space

4. Is a vector space

6. Not a vector space

8. Not a vector space

10. Not a vector space

12. Is a vector space

14. Not a subspace

16. Is a subspace

18. Is a subspace

20. Is a subspace

SECTION 6.2

2. $1 + x + 2x^2 = (1)(1 + x^2) + (1)(x + x^2)$

4. Independent

6. Dependent; $\begin{bmatrix} 1 & 0 \\ 1 & 2 \end{bmatrix} = (1)\begin{bmatrix} 1 & 2 \\ -1 & 0 \end{bmatrix} + (2)\begin{bmatrix} 0 & -1 \\ 1 & 1 \end{bmatrix}$

8. Independent

10. Independent

12. Dependent; $\begin{bmatrix} 0 & 1 \\ 0 & 1 \end{bmatrix} = (-1)\begin{bmatrix} 1 & 0 \\ 1 & 0 \end{bmatrix} + (1)\begin{bmatrix} 1 & 0 \\ 0 & 1 \end{bmatrix} + (1)\begin{bmatrix} 0 & 1 \\ 1 & 0 \end{bmatrix}$

14. Does not generate

16. Generates

18. Does not generate

20. Does not generate

22. Does not generate

24. Generates

26. Not a basis

28. Is a basis

30. Dependent, not a basis

32. Not a basis

34. Not a basis

36. Not a basis

38. 6

40. 2

42. 3

SECTION 6.3

2. (1, 2, 1, -1, 1, 2)

4. (-2, 3)

6. $x^2 + 4x - 5 = 2(x^2 + x - 1) + (-1)(x^2 - 2x + 3)$

8. Independent

10. Dependent; $\begin{bmatrix} 3 & -1 \\ -1 & 2 \end{bmatrix} = (5)\begin{bmatrix} 1 & 0 \\ 0 & 1 \end{bmatrix} + (-1)\begin{bmatrix} 2 & 1 \\ 1 & 3 \end{bmatrix}$

12. Dependent; $\begin{bmatrix} 0 & 1 \\ 0 & 1 \end{bmatrix} = (-1)\begin{bmatrix} 1 & 0 \\ 1 & 0 \end{bmatrix} + (1)\begin{bmatrix} 1 & 0 \\ 0 & 1 \end{bmatrix} + (1)\begin{bmatrix} 0 & 1 \\ 1 & 0 \end{bmatrix}$

14. $\{1,\ x + 1,\ x^2 + x, x^3 + x^2 + x\}$

16. The set consisting of the matrices:

$$\begin{bmatrix} 1 & 2 \\ 3 & 1 \end{bmatrix}, \begin{bmatrix} 1 & 3 \\ 2 & 1 \end{bmatrix}, \begin{bmatrix} 1 & 0 \\ 0 & 1 \end{bmatrix}, \begin{bmatrix} 0 & 1 \\ 0 & 0 \end{bmatrix}$$

18. $\{x^2 - x + 1,\ x^2 + 2x + 1,\ x - 2\}$

20. The set consisting of the matrices:

$$\begin{bmatrix} 0 & 0 & 0 \\ 0 & -1 & 0 \\ 0 & 0 & 1 \end{bmatrix}, \begin{bmatrix} 2 & 0 & 0 \\ 0 & 1 & 0 \\ 0 & 0 & 1 \end{bmatrix}, \begin{bmatrix} 1 & 0 & 0 \\ 0 & 1 & 0 \\ 0 & 0 & 1 \end{bmatrix}$$

SECTION 6.4

2. Not an inner product; property (d)

4. Is an inner product

6. Is an inner product

8. Orthogonal, but not orthonormal

10. Not orthogonal

12. Orthonormal

14. Orthogonal, but not orthonormal

16. Not orthogonal

18. $\mathbf{x} = (1/3)(1/3, -2/3, 2/3) + (4/3)(-2/3, 1/3, 2/3) + (-1/3)(2/3, 2/3, 1/3)$

20. $\{(1/2, -1/2, 1/2, -1/2),\ (1/2, 1/2, 1/2, 1/2),\ (\sqrt{2}/2, 0, -\sqrt{2}/2, 0)\}$

22. $\{(\sqrt{2}/2,\ 0,\ -\sqrt{2}/2),\ (\sqrt{2}/2, 0, \sqrt{2}/2)\}$

24. $\{(\sqrt{7}/7)(-1, -1, 2, 1),\ (\sqrt{3}/3)(1, 0, 1, -1),\ (\sqrt{35}/35)(3, 3, 1, 4)\}$

26. $\{(\sqrt{3}/3)(1, 1, 1, 0),\ (\sqrt{33}/33)(2, -4, 2, -3),\ (\sqrt{22}/22)(1, -2, 1, 4),\ (\sqrt{2}/2)(1, 0, 1, 0)\}$

28. $\{1,\ \sqrt{3}(2x - 1),\ \sqrt{5}(6x^2 - 6x + 1)\}$

30. $\{1,\ (e^{1-x} - 1 + e)/(\sqrt{7e^2 - 12e + 5})\}$

SECTION 6.5

2. $p(x) = (18/5)x$

4. $p(x) = (2/9)(11 - 4x)$

6. $p(x) = -(12/5)x$

8. $p(x) = (487/162) - (128/81)x + (14/81)x^2$

10. $p(x) = 3x + x^3$

12. $p(x) = \pi^2/3 - 4\cos x$

14. $p(x) = (e^{\pi} - e^{-\pi})[\sqrt{2} - (1/2)\cos x + (1/2)\sin x]$

16. $p(x) = 0$

REVIEW EXERCISES

2. Is a vector space

4. Not a vector space; properties (a), (e)

6. Is a vector space

8. Dependent; $\begin{bmatrix} 4 & -4 \\ 0 & 1 \end{bmatrix} = (1)\begin{bmatrix} 2 & 0 \\ 0 & 1 \end{bmatrix} + (2)\begin{bmatrix} 1 & -2 \\ 0 & 0 \end{bmatrix}$

10. Dependent; $\begin{bmatrix} -1 & 2 \\ 3 & -1 \end{bmatrix} = (2)\begin{bmatrix} -1 & 1 \\ 2 & -1 \end{bmatrix} + (1)\begin{bmatrix} 1 & 0 \\ -1 & 1 \end{bmatrix}$

12. Not orthogonal

14. Not orthogonal

16. Not orthogonal

18. Is an inner product

20. Is an inner product

22. An orthonormal basis is given by the two vectors:

$$\frac{2e^{x}}{\sqrt{e^{2}-1}}, \quad \frac{2(e^{2}-1)}{\sqrt{e^{2}-6+e^{-2}}}\left(e^{-x}-\frac{2}{\sqrt{e^{2}-1}}e^{x}\right)$$

CHAPTER 7

Linear Transformations

Chapter Overview

The topics covered in this chapter are:

1. Definition of a *linear transformation* from one vector space to another ($T: \mathbf{V} \to \mathbf{W}$); examples, including identity and zero transformations, rotations, contractions, dilations, and reflections. [Section 7.1]

2. Some basic properties of linear transformations, such as $T(\mathbf{0}) = \mathbf{0}$. [7.1]

3. Sums, differences, negatives, scalar multiples, compositions (products), and powers of linear transformations; properties of these operations. [7.2]

4. Matrix of a linear transformation from $\mathbf{R}^n$ to $\mathbf{R}^m$; matrices associated with sums, scalar multiples, and compositions. [7.2]

5. *Kernel* and *image* of a linear transformation; the fact that each is a subspace of its respective vector space. [7.3]

6. *Nullity* (dimension of kernel) and *rank* (dimension of image) of a linear transformation; if $T: \mathbf{V} \to \mathbf{W}$ is a linear transformation and $\dim(\mathbf{V})$ is n, $\dim(\mathbf{W})$ is m, then: [7.3]

   - nullity + rank = n (proved only for linear transformations from $\mathbf{R}^n$ to $\mathbf{R}^m$).
   - T is one-to-one if nullity = 0; T is onto if rank = m.

7. *Inverse* of a linear transformation from $\mathbf{R}^n$ to $\mathbf{R}^m$. [7.3]

8. Application of linear transformations to differential equations, including: [7.4]

   - Linear differential equations of order n;
   - Solving homogeneous linear differential equations; Wronskians;
   - Form of the solution to a nonhomogeneous linear differential equation;
   - Initial value problems; existence and uniqueness of solutions.

9. Matrix of a linear transformation between arbitrary (finite-dimensional) vector spaces; discussion of kernel and image in this more general context. [7.5]

10. Change of basis matrix; similarity transformations. [7.6]

Remarks

Because many students do not have a clear idea of what a function really is, the idea of a linear transformation is more difficult than it should be. The simplest case of linear transformations from $\mathbf{R}^n$ to $\mathbf{R}^m$ defined in terms of matrix multiplication can (and should) become mechanical, but the underlying concepts are more difficult, yet very important. Some practice with non-matrix examples will help develop this understanding. Rotation, reflection, and contraction without using matrices provide ways of motivating the ideas as well as the convenience of the matrix representation. Examples 7 and 9 of Section 7.1, for example, demonstrate these ideas without losing sight of the concept in the mechanics.

In Section 7.2, the algebra of linear transformations is presented and the important connection between transformations and the matrices that represent them is introduced. (To ease the learning process, only matrices associated with linear transformations from $\mathbf{R}^n$ to $\mathbf{R}^m$ are developed in this section,) It is, of course, Theorem 6 which finally really explains why matrix multiplication is defined as it is — in order to represent composition of the underlying linear transformations. This fact should be observed for the benefit of those in the class who can appreciate it.

Section 7.3 introduces two important subspaces associated with a linear transformation, the kernel in the domain space and the image in the range space. In the case of Euclidean spaces (which are emphasized in this section), both subspaces turn out to be nothing more than old concepts in a new setting; namely, solution of a homogeneous system of linear equations and the linear span of the columns of a matrix. Making this connection will help some students from becoming too lost in the abstraction.

Using matrix methods when the given vector spaces are not Euclidean m-spaces is developed in Section 7.5. This is one of those places in mathematics where formalizing a rather simple idea makes it much more difficult. A couple of examples similar to those in the text should help to communicate the concept. The formal description of the process, Theorems 1 - 3, is quite another matter.

If the formalization of the subject matter of Section 7.5 gave your students trouble, then Section 7.6 will probably be much too difficult for them. We have tried to minimize the difficulty by staying within the context of Euclidean space and using only one basis instead of two, but this material is still far from easy. Again, the idea is less difficult than the formalization of it, but this time it is not possible to entirely escape from the formality.

Solutions to Self-Test

1. The function L is not a linear transformation; it is neither additive nor do scalars factor out. An example of either is sufficient; the easiest way to show that L is not a linear transformation, however is to notice that

$$\mathrm{L}(\mathbf{0}) = \mathrm{L}(0, 0) = (0, 0, 1)$$

which is not the zero vector in $\mathbf{R}^3$.

2. The function K is a linear transformation. For two vectors

$$\mathbf{p}(x) = a + bx + cx^2 \quad \text{and} \quad \mathbf{q}(x) = d + ex + fx^2$$

and scalar s, we must show that

$$\mathrm{K}(s\mathbf{p}) = s\mathrm{K}(\mathbf{p}) \quad \text{and} \quad \mathrm{K}(\mathbf{p}+\mathbf{q}) = \mathrm{K}(\mathbf{p}) + \mathrm{K}(\mathbf{q}).$$

$$\begin{aligned} \text{Now, } \mathrm{K}(s\mathbf{p}) &= \mathrm{K}(s(a + bx + cx^2)) \\ &= \mathrm{K}(sa + sbx + scx^2) \\ &= \begin{bmatrix} 0 & sa - sb \\ sa + sb & sc \end{bmatrix} \\ &= s\begin{bmatrix} 0 & a - b \\ a + b & c \end{bmatrix} \\ &= s\mathrm{K}(\mathbf{p}) \end{aligned}$$

$$\begin{aligned} \text{Also, } \mathrm{K}(\mathbf{p}+\mathbf{q}) &= \mathrm{K}((a + bx + cx^2) + (d + ex + fx^2)) \\ &= \mathrm{K}((a + d) + (b + e)x + (c + f)x^2) \end{aligned}$$

$$= \begin{bmatrix} 0 & (a+d) - (b+e) \\ (a+d) + (b+e) & c+f \end{bmatrix}$$

$$= \begin{bmatrix} 0 & (a-b) + (d-e) \\ (a+b) + (d+e) & c+f \end{bmatrix}$$

$$= \begin{bmatrix} 0 & (a-b) \\ (a+b) & c \end{bmatrix} + \begin{bmatrix} 0 & (d-e) \\ (d+e) & f \end{bmatrix}$$

$$= \mathrm{K}(a + bx + cx^2) + \mathrm{K}(d + ex + fx^2)$$
$$= \mathrm{K}(\mathbf{p}) + \mathrm{K}(\mathbf{q})$$

3. $\mathrm{T}^2(\mathbf{x}) = \mathrm{T}(\mathrm{T}(\mathbf{x}))$
$= \mathrm{T}(\mathrm{T}(x, y, z)) = \mathrm{T}(z, x + y, 0)$
$= (0, z + (x + y), 0) = (0, x + y + z, 0)$

4. In fact, in this situation the matrix can be directly obtained from the coefficients, but this fact was never formally presented. Following the theorem that was given, we compute $\mathrm{T}(\mathbf{e}_i)$ for each $i = 1, 2, 3$, and the resulting vectors are the columns of the desired matrix.

$$\mathrm{T}(\mathbf{e}_1) = \mathrm{T}(1, 0, 0) = (0, 1, 0)$$
$$\mathrm{T}(\mathbf{e}_2) = \mathrm{T}(0, 1, 0) = (0, 1, 0)$$
$$\mathrm{T}(\mathbf{e}_3) = \mathrm{T}(0, 0, 1) = (1, 0, 0)$$

Thus, the matrix that represents T is

$$\mathrm{A} = \begin{bmatrix} 0 & 0 & 1 \\ 1 & 1 & 0 \\ 0 & 0 & 0 \end{bmatrix}$$

5. (a) The kernel of T is the same set as the solutions to the homogeneous system $\mathrm{A}\mathbf{x} = \mathbf{0}$, where A is the matrix given in the solution to Exercise 4. Solving this system by the usual row-reduction process, we see that the solution is the set of all vectors of the form $(-s, s, 0)$. A basis can be found by setting $s = 1$. That is, a basis is the set containing the single vector $(-1, 1, 0)$.

(b) T is not one-to-one because Ker(T) contains nonzero vectors.

6. **(a)** The range of T is the span of the columns of the matrix A of Exercise 4. A basis for this subspace of $\mathbf{R}^2$ is given by the columns of A that correspond to those columns in its row-reduced echelon form which make up a distinct set of elementary vectors. Since the row-reduced form is

$$B = \begin{bmatrix} 1 & 1 & 0 \\ 0 & 0 & 1 \\ 0 & 0 & 0 \end{bmatrix},$$

one such set is $\{\mathbf{a}^1, \mathbf{a}^3\}$, which corresponds to the first and third columns of B; another is $\{\mathbf{a}^2, \mathbf{a}^3\}$.

(b) T is not onto because the dimension of Im(T) is 2 in $\mathbf{R}^3$, which is of dimension 3. Alternatively, from Exercise 5, we see that T cannot be onto because it is a linear operator that is not one-to-one.

7. **(a)** $L(x, y) = A\mathbf{x} = (2x + 5y,\ x + 3y)$

(b) The inverse of the matrix A is

$$A^{-1} = \begin{bmatrix} 3 & -5 \\ -1 & 2 \end{bmatrix}$$

Then, the formula for $L^{-1}(\mathbf{x})$ can be obtained by multiplying the matrix A^{-1} by the column vector $[x\ \ y\ \ z]^T$; that is,

$$L^{-1}(\mathbf{x}) = A^{-1}\mathbf{x} = (3x - 5y,\ -x + 2y).$$

8. For each basis element $\mathbf{b}$ of $\mathbf{P}_2$, we express S($\mathbf{b}$) as a linear combination of the basis elements of $\mathbf{R}^4$ and use the vectors of coefficients as the columns of the matrix. Since these are just the vectors S($\mathbf{b}$), the work is a little simpler than it otherwise might be:

$$S(1) = (0, 1, 0, 4) = 0\mathbf{e}_1 + 1\mathbf{e}_2 + 0\mathbf{e}_3 + 4\mathbf{e}_4$$
$$S(x) = (0, -1, 0, 0) = 0\mathbf{e}_1 - 1\mathbf{e}_2 + 0\mathbf{e}_3 + 0\mathbf{e}_4$$
$$S(x^2) = (1, 0, 0, 0) = 1\mathbf{e}_1 + 0\mathbf{e}_2 + 0\mathbf{e}_3 + 0\mathbf{e}_4$$

Thus, the matrix is given by

$$B = \begin{bmatrix} 0 & 0 & 1 \\ 1 & -1 & 0 \\ 0 & 0 & 0 \\ 4 & 0 & 0 \end{bmatrix}$$

9. **(a)** The row-reduced echelon form for B is just three distinct elementary columns, so Ker(S) = $\{\mathbf{0}\}$ and Ker(S) has dimension 0. Thus, S is one-to-one.

(b) From the row-reduced form for B, it follows that the columns of B are linearly independent. Thus, the dimension of Im(S) is 3. Because $\mathbf{R}^4$ has dimension 4, S is not onto.

10. The change of basis matrix is the inverse of the matrix of columns of the vectors in the basis. Computing that matrix we have the change of basis matrix:

$$A^{-1} = \begin{bmatrix} 1 & -1 & 0 \\ 0 & 1 & -1 \\ 0 & 0 & 1 \end{bmatrix}$$

Solutions to Odd-Numbered Review Exercises

1. Let $\mathbf{x}$, $\mathbf{x}_1$, and $\mathbf{x}_2$ be elements of $\mathbf{R}^3$ and let c be a scalar. We need to verify that $S(c\mathbf{x}) = cS(\mathbf{x})$ and that $S(\mathbf{x}_1 + \mathbf{x}_2) = S(\mathbf{x}_1) + S(\mathbf{x}_2)$.

$$\begin{aligned} S(c\mathbf{x}) &= S(c(x, y, z)) \\ &= S(cx, cy, cz) \\ &= (cx - 2cy,\ cx + cy + cz) \\ &= c(x - 2y,\ x + y + z) \\ &= cS(x, y, z) = cS(\mathbf{x}) \end{aligned}$$

Thus, scalars factor out. To show that S is additive:

$$\begin{aligned} S(\mathbf{x}_1 + \mathbf{x}_2) &= S((x_1, y_1, z_1) + (x_2, y_2, z_2)) \\ &= S(x_1 + x_2,\ y_1 + y_2,\ z_1 + z_2) \\ &= ((x_1 + x_2) - 2(y_1 + y_2),\ (x_1 + x_2) + (y_1 + y_2) + (z_1 + z_2)) \\ &= ((x_1 - 2y_1) + (x_2 - 2y_2),\ (x_1 + y_1 + z_1) + (x_2 + y_2 + z_2)) \\ &= (x_1 - 2y_1,\ x_1 + y_1 + z_1) + (x_2 - 2y_2,\ x_2 + y_2 + z_2) \\ &= S(x_1, y_1, z_1) + S(x_2, y_2, z_2) \end{aligned}$$

$$= S(\mathbf{x}_1) + S(\mathbf{x}_2)$$

3. Specific counterexamples abound and any one is sufficient. The example given in the "odd-numbered answers" in the text shows that S is not additive. For the sake of variety, here we choose an example to show that scalars do not factor out. Let $c = 2$ and $\mathbf{x} = (1, 1)$. Then:

$$S(2(1, 1)) = S(2, 2) = (2 + 2(2), 2(2)) = (6, 4).$$

But,

$$2S(1, 1) = 2(1 + (2(1), 1(1)) = 2(3, 1) = (6, 2).$$

5. The range of D is $\mathbf{R}^3$ while the domain of F is $\mathbf{R}^2$. Since these spaces are different, FD is not defined.

7.
$$\begin{aligned} E^2(x, y, z) &= E(E(x, y, z)) \\ &= E(x,\ x + y + z,\ 2x + y + z) \\ &= (x,\ x + (x + y + z) + (2x + y + z),\ 2x + (x + y + z) + (2x + y + z)) \\ &- (x,\ 4x + 2y + 2z,\ 5x + 2y + 2z) \end{aligned}$$

9. Apply D to each basis vector and form the matrix of columns of these vectors:

$$D(\mathbf{e}_1) = D(1, 0) = (1 + 0, 1 - 2(0), 1) = (1, 1, 1)$$

$$D(\mathbf{e}_2) = D(0, 1) = (0 + 1, 0 - 2(1), 0) = (1, -2, 0)$$

From Theorem 4 of Section 7.2, the matrix A that represents D is given by

$$\begin{bmatrix} 1 & 1 \\ 1 & -2 \\ 1 & 0 \end{bmatrix}$$

11. From Theorem 3 of Section 7.3, we need to show that $\text{Ker}(D) = \{\mathbf{0}\}$ to conclude that D is one-to-one. By Theorem 2, this is equivalent to showing that $A\mathbf{x} = \mathbf{0}$ has only the trivial solution, where A is given in Exercise 9. Since the row-reduced echelon form of A augmented by $\mathbf{0}$ is

$$\left[\begin{array}{cc|c} 1 & 0 & 0 \\ 0 & 1 & 0 \\ 0 & 0 & 0 \end{array}\right]$$

the only solution is $\mathbf{x} = \mathbf{0}$, so D is one-to-one.

13. Following the argument of Exercise 11, we need only verify that $A\mathbf{x} = \mathbf{0}$ has multiple solutions to conclude that E is not one-to-one, where

$$A = \begin{bmatrix} 1 & 0 & 0 \\ 1 & 1 & 1 \\ 2 & 1 & 1 \end{bmatrix}$$

Since A is square, $A\mathbf{x} = \mathbf{0}$ will have more than one solution if and only if $\det(A) = 0$. Computing this determinant, we obtain $\det(A) = 0$ and therefore E is not one-to-one.

15. $\text{Ker}(D) = \{\mathbf{0}\}$; see the solution to Exercise 11.

17. Letting

$$B = \{1, x, x^2, x^3\} \quad \text{and} \quad B' = \{E^{11}, E^{12}, E^{21}, E^{22}\}$$

be the ordered bases of $\mathbf{P}_3$ and $\mathbf{M}^{2,2}$, respectively, express the image under T of each element of the first as a linear combination of the second:

$$T(1) = \begin{bmatrix} 1 & 0 \\ 0 & 2 \end{bmatrix} = 1E^{11} + 2E^{22} \rightarrow (1, 0, 0, 2)$$

$$T(x) = \begin{bmatrix} 0 & 1 \\ 0 & 1 \end{bmatrix} = 1E^{12} + 1E^{22} \rightarrow (0, 1, 0, 1)$$

$$T(x^2) = \begin{bmatrix} 1 & 0 \\ 2 & 0 \end{bmatrix} = 1E^{11} + 2E^{21} \rightarrow (1, 0, 2, 0)$$

$$T(x^3) = \begin{bmatrix} 0 & -1 \\ 0 & -1 \end{bmatrix} = -1E^{12} - 1E^{22} \rightarrow (0, -1, 0, -1)$$

The desired matrix is the matrix with these vectors as columns:

$$\begin{bmatrix} 1 & 0 & 1 & 0 \\ 0 & 1 & 0 & -1 \\ 0 & 0 & 2 & 0 \\ 2 & 1 & 0 & -1 \end{bmatrix}$$

19. T is not onto. The easiest way to verify this fact is to note that for the matrix A in the solution to Exercise 17, det(A) = 0. By Theorem 4d of Section 7.5, T is not onto.

21. Following the procedure of Section 7.5 with $\mathbf{B} = \mathbf{B}'$, express the image under E of each basis element of $\mathbf{B}$ as a linear combination of the basis $\mathbf{B}'$:

$$\begin{aligned} E(1, 1, 0) &= (1,\ 1+1+0,\ 2(1)+1+0) = (1, 2, 3) \\ &= 0(1, 1, 0) + 1(1, 0, 1) + 2(0, 1, 1) \\ &\rightarrow (0, 1, 2) \end{aligned}$$

$$\begin{aligned} E(1, 0, 1) &= (1,\ 1+0+1,\ 2(1)+0+1) = (1, 2, 3) \\ &= 0(1, 1, 0) + 1(1, 0, 1) + 2(0, 1, 1) \\ &\rightarrow (0, 1, 2) \end{aligned}$$

$$\begin{aligned} E(0, 1, 1) &= (0,\ 0+1+1,\ 2(0)+1+1) = (0, 2, 2) \\ &= 0(1, 1, 0) + 0(1, 0, 1) + 2(0, 1, 1) \\ &\rightarrow (0, 0, 2) \end{aligned}$$

The desired matrix is the one with these vectors as columns:

$$A_B = \begin{bmatrix} 0 & 0 & 0 \\ 1 & 1 & 0 \\ 2 & 2 & 2 \end{bmatrix}$$

Answers to Even-Numbered Exercises

SECTION 7.1

2. To show that $T(x, y) = (2x, x + y, x - 2y)$ is linear we need to verify both *i.* and *ii.* of the definition. For any vector (a, b) in $\mathbf{R}^2$ and scalar c, we have

$$\begin{aligned} T(c(a, b)) &= T((ca, cb)) \\ &= (2(ca),\ ca + cb,\ ca - 2(cb)) \\ &= c(2a,\ a + b,\ a - 2b) = cT(a,b). \end{aligned}$$

Thus, scalars factor out; that is, condition *i.* is satisfied. Let (a, b) and (c, d) by any vectors in $\mathbf{R}^2$. Then:

$$\begin{aligned} T((a, b) + (c, d)) &= T((a + c, b + d)) \\ &= (2(a + c),\ (a + c) + (b + d),\ (a + c) - 2(b + d)) \\ &= (2a + 2c,\ (a + b) + (c + d),\ (a - 2b) + (c - 2d)) \\ &= (2a, a + b, a - 2b) + (2c, c + d, c - 2d) \end{aligned}$$

$$= \mathrm{T}(a, b) + \mathrm{T}(c, d)$$

Thus T is additive as well, so T is a linear transformation.

4. For any vectors $\mathbf{x}$, $\mathbf{y}$ in $\mathbf{R}^n$ and scalar c, we have

$$\mathrm{I}(c\mathbf{x}) = c\mathbf{x} = c\mathrm{I}(\mathbf{x}),$$

so scalars factor out. Also,

$$\mathrm{I}(\mathbf{x}+\mathbf{y}) = \mathbf{x}+\mathbf{y} = \mathrm{I}(\mathbf{x}) + \mathrm{I}(\mathbf{y}),$$

so I is additive. Thus, I is a linear transformation.

6. For arbitrary polynomials of degree 3,

$$\mathbf{p}(x) = a_0 + a_1x + a_2x^2 + a_3x^3 \quad \text{and} \quad \mathbf{q}(x) = b_0 + b_1x + b_2x^2 + b_3x^3,$$

and scalar c, we have both

$$\begin{aligned} \mathrm{T}(c\mathbf{p}) &= \mathrm{T}(c(a_0 + a_1x + a_2x^2 + a_3x^3)) \\ &= \mathrm{T}(ca_0 + ca_1x + ca_2x^2 + ca_3x^3) \\ &= (ca_0 + ca_2) - (ca_1 + 2ca_3)x^2 \\ &= c[(a_0 + a_2) - (a_1 + 2a_3)x^2] \\ &= c\mathrm{T}(a_0 + a_1x + a_2x^2 + a_3x^3) \\ &= c\mathrm{T}(\mathbf{p}) \end{aligned}$$

and

$$\begin{aligned} \mathrm{T}(\mathbf{p}+\mathbf{q}) &= \mathrm{T}((a_0 + b_0) + (a_1 + b_1)x + (a_2 + b_2)x^2 + (a_3 + b_3)x^3) \\ &= [(a_0 + b_0) + (a_2 + b_2)] - [(a_1 + b_1) + 2(a_3 + b_3)]x^2 \\ &= [a_0 + a_2 - (a_1 + 2a_3)x^2)] + [b_0 + b_2 - (b_1 + 2b_3)x^2] \\ &= \mathrm{T}(\mathbf{p}) + \mathrm{T}(\mathbf{q}). \end{aligned}$$

Thus, T is a linear transformation.

8. Similar to Exercise 6.

10. Not linear

12. Not linear

14. $n = m = 3$, $\mathrm{T}(\mathbf{x}_0) = (7, 16, 3)$

16. $n = m = 3$, $\mathrm{T}(\mathbf{x}_0) = (y, x, z)$

22. $\begin{bmatrix} 0 & 1 \\ -1 & 0 \end{bmatrix}$

24. $\begin{bmatrix} 1 & 0 \\ 0 & -1 \end{bmatrix}$

32. The following matrix must be raised to the 720th power:

$$\sqrt[720]{1/1000}\begin{bmatrix} \cos 1° & -\sin 1° \\ \sin 1° & \cos 1° \end{bmatrix}$$

SECTION 7.2

2. $TL(x, y, z) = (5x,\ 3x + y,\ x - 3y,\ -5y)$

4. $(3K - 2S)(x, y, z) = (x + z,\ x + 2y,\ 2x + y + z)$

6. $GS(x, y, z) = \begin{bmatrix} y & x \\ y - z & -z \end{bmatrix}$

8. Not defined

10. $(2S^3 - S^2 + 3S - 4I)(x, y, z) = (-5x + 5z,\ 0,\ 5x - 5z)$

12. $HGK(a, b, c) = (2a + 2b + c) + (3a + b)x - ax^2$

14. $\begin{bmatrix} 2 & -1 & 0 \\ 1 & 2 & 0 \end{bmatrix}$

16. $\begin{bmatrix} 2 & 1 \\ 1 & 1 \\ 1 & -1 \\ 1 & -2 \end{bmatrix}$

18. $\begin{bmatrix} 5 & 0 & 0 \\ 3 & 1 & 0 \\ 1 & -3 & -5 \end{bmatrix}$

20. $\begin{bmatrix} 1 & 0 & 1 \\ 1 & 2 & 0 \\ 2 & 1 & 1 \end{bmatrix}$

SECTION 7.3

2. $\mathbf{v}_1$ is not, $\mathbf{v}_2$ is

4. $\mathbf{v}_1$ is, $\mathbf{v}_2$ is

6. $\mathbf{v}_1$ is, $\mathbf{v}_2$ is

8. $\mathbf{w}_1$ is, $\mathbf{w}_2$ is not

10. $\mathbf{w}_1$ is, $\mathbf{w}_2$ is

12. $\mathbf{w}_1$ is, $\mathbf{w}_2$ is

14. nullity: 0; basis: {} (the empty set)

16. nullity: 3; basis: {(-2, 1, 0, 0), (-3, 0, 1, 0), (-4, 0, 0, 1)}

18. nullity: 3; basis consists of the following vectors:

$$\begin{bmatrix} 0 & 1 & 0 \\ 0 & 0 & 0 \end{bmatrix}, \begin{bmatrix} -1 & 0 & 1 \\ 0 & 0 & 0 \end{bmatrix}, \begin{bmatrix} 0 & 0 & 0 \\ 1 & 1 & 1 \end{bmatrix}$$

20. rank: 3; basis: $\{(1, 2, -1, 1), (1, 0, 1, 1), (0, 3, 1, 2)\}$

22. rank: 1; basis: $\{1\}$

24. rank: 3; basis: $\{x^2, x, 1\}$

30. not invertible

32. $T^{-1}(\mathbf{x}) = B\mathbf{x}$, where

$$B = \begin{bmatrix} 0 & -8 & -2 & -3 \\ 0 & -4 & -1 & -2 \\ 1 & 0 & 0 & 1 \\ 0 & 3 & 1 & 1 \end{bmatrix}$$

34. T^{θ} is its own inverse.

SECTION 7.4

2. $y = c_1 e^{-3x} + c_2 e^{x}$

4. $y = c_1 e^{(2+\sqrt{3})} + c_2 e^{(2-\sqrt{3})}$

6. $u = c_1 e^{x/2} + c_2 x e^{x/2}$

8. $s = c_1 \cos(3t) + c_2 \sin(3t)$

10. $y = e^{x}(c_1 \cos(\sqrt{3}x) + c_2 \sin(\sqrt{3}x))$

12. $u = c_1 + c_2 e^{2x} + c_3 x e^{2x}$

14. $s = c_1 + c_2 t + c_3 \cos t + c_4 \sin t$

16. $s = c_1 \cos t + c_2 \sin t + c_3 t \cos t + c_4 t \sin t$

18. $y = c_1 e^{x} + c_2 e^{-x} - x^2 - 3$

20. $u = c_1 + c_2 e^{-x} + \sin x - \cos x$

22. $y = 1 + (1/e) e^{x}$

24. $s(t) = -2 + t$

26. $u = 2 + \sin x$

28. $s(t) = -2 \cos 3t$

30. $\theta(t) = (\pi/6) \cos 4t$

SECTION 7.5

2. $\begin{bmatrix} 1 & 0 & 1 & 0 \\ 0 & 0 & 0 & 0 \\ 0 & -1 & 0 & -2 \\ 0 & 0 & 0 & 0 \end{bmatrix}$

4. $\begin{bmatrix} 1/2 & 0 & 0 \\ 1/2 & 0 & 1 \\ 0 & 1 & 1 \\ 0 & 1 & 1 \end{bmatrix}$

6. Ker(T) = $\{-a - 2bx + ax^2 + bx^3 \mid a, b \text{ in } \mathbf{R}\}$; not one-to-one.

8. Ker(T) = $\{\mathbf{0}\}$; is one-to-one

10. Im(T) = $\{a + bx^2 \mid a, b \text{ in } \mathbf{R}\}$; not onto

12. Im(T) = $\mathbf{P}_2$; onto

14. D(**P**) = $2x - 9x^2 + 8x^3$

SECTION 7.6

2. $\frac{1}{2}\begin{bmatrix} 1 & -1 \\ 0 & 2 \end{bmatrix}$, (2, -1)

4. $\frac{1}{6}\begin{bmatrix} -1 & 5 & -3 \\ -4 & 2 & 0 \\ 3 & -3 & 3 \end{bmatrix}$, (1/2, -1, 1/2)

6. $\frac{1}{20}\begin{bmatrix} -6 & 9 & -2 & 3 \\ 6 & 11 & 2 & -3 \\ 2 & -13 & -6 & 9 \\ -2 & 3 & 6 & 1 \end{bmatrix}$, (1/10, 9/10, 3/10, 7/10)

8. $\begin{bmatrix} 1 & 3/2 \\ 0 & -1 \end{bmatrix}$

10. $\frac{1}{3}\begin{bmatrix} -8 & -7 & -6 \\ -11 & -13 & -21 \\ 15 & 12 & 21 \end{bmatrix}$

12. $\frac{1}{6}\begin{bmatrix} 17 & 16 & 33 \\ 8 & 16 & 12 \\ -9 & -18 & -21 \end{bmatrix}$

14. $\frac{1}{20}\begin{bmatrix} 7 & 20 & 12 & -49 \\ -27 & 60 & 28 & -11 \\ 1 & -80 & -64 & -7 \\ 9 & 40 & 24 & 17 \end{bmatrix}$

REVIEW EXERCISES

2. Verify the two properties *in general.*

4. Produce a *specific* counterexample, such as:

T((-1)(0, 1)) = T(0, -1) = (0, -1, 1) but (-1)T(0, 1) = (-1)(0, 1, 1) = (0, -1, -1)

6. (D - 2F)(x, y) = $(-x + y,\ -3x - 4y,\ -3x + 2y)$

8. Not defined

10. $\begin{bmatrix} 1 & 0 & 0 \\ 1 & 1 & 1 \\ 2 & 1 & 1 \end{bmatrix}$

12. No, rank(D) = 3 < 2.

14. No, rank(E) = 2 < 3

16. $\{(0, -s, s) \mid s \text{ in } \mathbf{R}\}$

18. No, $T(x + x^3) = 0$

20. $\{ax + ax^2 \mid a \text{ in } \mathbf{R}\}$

22. $\begin{bmatrix} 0 & 0 & 0 \\ 1 & 1 & 0 \\ 2 & 2 & 2 \end{bmatrix}$

CHAPTER 8

Eigenvalues and Eigenvectors

Chapter Overview

The topics covered in this chapter are:

1. *Eigenvalues* and *eigenvectors* of a linear operator and a square matrix. [Section 8.1]
2. Calculating the eigenvalues of a square matrix; the *characteristic polynomial* and *characteristic equation*. [8.1]
3. Calculating the eigenvectors and *eigenspaces* corresponding to a given eigenvalue. [8.1]
4. Finding the eigenvalues of lower triangular, upper triangular, and diagonal matrices. [8.1]
5. Markov chains, transition matrices, and equilibrium vectors. [8.2]
6. Procedure for *diagonalizing* a square matrix. [8.3]
7. The fact that an $n \times n$ matrix is diagonalizable: [8.3]
   - If and only if it has n linearly independent eigenvectors.
   - If it has n distinct eigenvalues.
8. *Symmetric matrices*; eigenvectors corresponding to distinct eigenvalues of a symmetric matrix are orthogonal. [8.3]
9. *Orthogonal matrices* ($P^{-1} = P^T$); the columns of P form an *orthonormal set*. [8.3]

10. If A is a symmetric matrix, then there exists an orthogonal matrix P such that $P^{-1}AP$ is diagonal (A is *orthogonally diagonalizable*). [8.3]
11. Types of *quadric surfaces*; using orthogonal diagonalization (and completing the square) to reduce the equation of certain quadric surfaces to standard form. [8.4]
12. Procedure for factoring a square matrix A as A = QR, where Q is an orthogonal matrix and R is an upper triangular matrix. [8.5]
13. The QR algorithm for approximating the eigenvalues of a square matrix. [8.5]
14. Complex *m*-space ($\mathbf{C}^m$), matrices with complex entries, complex eigenvalues and eigenvectors. [8.6]
15. Diagonalization of complex matrices; hermitian matrices ($A = A^H$); A is a hermitian matrix if and only if there exists a *unitary matrix* P ($P^{-1} = P^H$) such that $P^{-1}AP$ is a diagonal matrix with real entries. [8.6]

Remarks

This chapter contains a lot of material, and time (or the lack of it) may be a major consideration this late in the course. At the very least, you should try to cover the definitions and computational methods of Section 8.1. If you are really pressed for time, the material at the very beginning of the section dealing with eigenvalues of linear transformations can be omitted (together with the corresponding exercises). This strategy will not cause much of a problem as the chapter revolves around the eigenvalues of a square matrix.

Ideally, at least one nontrivial application should be presented as well. Be aware that Markov Chains (Section 8.2) is much easier to cover than Quadric Surfaces (Section 8.4). The latter requires material on orthogonally diagonalizing a matrix, which in turn uses the Gram-Schmidt process of Section 6.4.

The material on QR factorization of a matrix (Section 8.5) may be of interest to those instructors who want to at least hint at the use of numerical methods in linear algebra. Such instructors will presumably have already covered LU factorization (Section 3.6), which further motivates this material. If you do cover Section 8.5, be aware that:

- The definition of orthogonal matrix is given in Section 8.2, but that's the *only* material needed from that section.
- The Gram-Schmidt process (Section 6.4) is used in QR factorization.
- This text covers application of QR factorization to approximating

eigenvalues, but not to solving linear systems.

The last section of Chapter 8 introduces complex scalars and briefly discusses operations on complex m-vectors and matrices with complex entries. Most of the section is devoted to complex eigenvalues and eigenvectors, especially as they relate to the diagonalization of a matrix. This section might be of interest in those courses designed for engineers and science majors, for whom the complex field is important, but who may not be required to take a more advanced course in linear algebra.

Solutions to Self-Test

1. To find the eigenvalues of T, we need to find all numbers λ for which there is a nonzero vector $\mathbf{x} = (x, y)$ with $T(\mathbf{x}) = \lambda\mathbf{x}$; that is, $(x, 0) = \lambda(x, y)$ for some nonzero (x, y). This yields the linear system

$$\begin{aligned} x &= \lambda x \\ 0 &= \lambda y \end{aligned}$$

with at least one of x or y nonzero.

If $\lambda = 0$, then y is unrestricted, but x must be 0. Thus, $\lambda = 0$ is an eigenvalue with corresponding eigenvectors $(0, s)$ for $s \neq 0$ in **R**.

If $\lambda \neq 0$, then y must be 0, so nonzero (x, y) with $y = 0$ forces x to be nonzero. Dividing both sides of the first equation by x we have $\lambda = 1$ as the only nonzero eigenvalue and the corresponding eigenvectors are $(s, 0)$ for $s \neq 0$ in **R**.

Note: This problem can be done more mechanically by using the matrix A that represents T and solving the equation $\det(A - \lambda I) = 0$. The approach given above demonstrates a solution "from first principles."

2. **(a)** Simplify $\det(A - \lambda I) = 0$ for the given matrix A:

$$0 = \det\begin{bmatrix} 1 - \lambda & 1 \\ 0 & 1 - \lambda \end{bmatrix} = (1 - \lambda)^2$$

(b) Solve the equation of part (a): $\lambda = 1$

(c) For $\lambda = 1$, we must solve $(A - \lambda I)\mathbf{x} = \mathbf{0}$, which has the augmented matrix:

$$\left[\begin{array}{cc|c} 0 & 1 & 0 \\ 0 & 0 & 0 \end{array}\right]$$

This matrix is already in row-reduced echelon form; all solutions are given by $(s, 0)$, s in **R**. A basis is obtained by setting $s = 1$: $\{(1, 0)\}$.

3. **(a)** Simplify $\det(A - \lambda I) = 0$ for the given matrix A:

$$0 = \det \begin{bmatrix} 2-\lambda & 2 & 2 \\ 2 & 2-\lambda & 2 \\ 2 & 2 & 2-\lambda \end{bmatrix} = \lambda^2(6 - \lambda)$$

(b) Solve the equation of part (a): $\lambda = 0, 6$

(c) For each eigenvalue, λ, of part (b), solve $(A - \lambda I)\mathbf{x} = \mathbf{0}$:

$$\lambda = 0: \quad \left[\begin{array}{ccc|c} 2 & 2 & 2 & 0 \\ 2 & 2 & 2 & 0 \\ 2 & 2 & 2 & 0 \end{array}\right] \rightarrow \left[\begin{array}{ccc|c} 1 & 1 & 1 & 0 \\ 0 & 0 & 0 & 0 \\ 0 & 0 & 0 & 0 \end{array}\right]$$

All solutions are given by $(-s - t, s, t)$, where s and t are in **R**. A basis is obtained by setting $s = 1$ and $t = 0$ for one vector and $s = 0$ and $t = 1$ for the other: $\{(-1, 1, 0), (-1, 0, 1)\}$

$$\lambda = 6: \quad \left[\begin{array}{ccc|c} -4 & 2 & 2 & 0 \\ 2 & -4 & 2 & 0 \\ 2 & 2 & -4 & 0 \end{array}\right] \rightarrow \left[\begin{array}{ccc|c} 1 & 0 & -1 & 0 \\ 0 & 1 & -1 & 0 \\ 0 & 0 & 0 & 0 \end{array}\right]$$

All solutions are given by (s, s, s), where s is in **R**. A basis is obtained by setting $s = 1$: $\{(1, 1, 1)\}$

4. **(a)** Simplify $\det(A - \lambda I) = 0$ for the given matrix A:

$$0 = \det \begin{bmatrix} 2-\lambda & -1 & 0 \\ -1 & 2-\lambda & -1 \\ 0 & -1 & 2-\lambda \end{bmatrix} = (2 - \lambda)(\lambda^2 - 4\lambda + 2)$$

(b) Solve the equation of part (a): $\lambda = 2, 2 \pm \sqrt{2}$

(c) For each eigenvalue, λ, of part (b), solve $(A - \lambda I)\mathbf{x} = \mathbf{0}$:

$$\lambda = 2: \quad \left[\begin{array}{rrr|r} 0 & -1 & 0 & 0 \\ -1 & 0 & -1 & 0 \\ 0 & -1 & 0 & 0 \end{array}\right] \rightarrow \left[\begin{array}{rrr|r} 1 & 0 & 1 & 0 \\ 0 & 1 & 0 & 0 \\ 0 & 0 & 0 & 0 \end{array}\right]$$

All solutions are given by $(-s, 0, s)$, where s is in **R**. A basis is obtained by setting $s = 1$: $\{(-1, 0, 1)\}$

$$\lambda = 2 + \sqrt{2}: \quad \left[\begin{array}{rrr|r} -\sqrt{2} & -1 & 0 & 0 \\ -1 & -\sqrt{2} & -1 & 0 \\ 0 & -1 & -\sqrt{2} & 0 \end{array}\right] \rightarrow \left[\begin{array}{rrr|r} 1 & 0 & -1 & 0 \\ 0 & 1 & \sqrt{2} & 0 \\ 0 & 0 & 0 & 0 \end{array}\right]$$

All solutions are given by $(s, -s\sqrt{2}, s)$, where s is in **R**. A basis is obtained by setting $s = 1$: $\{(1, -\sqrt{2}, 1)\}$

$$\lambda = 2 - \sqrt{2}: \quad \left[\begin{array}{rrr|r} \sqrt{2} & -1 & 0 & 0 \\ -1 & \sqrt{2} & -1 & 0 \\ 0 & -1 & \sqrt{2} & 0 \end{array}\right] \rightarrow \left[\begin{array}{rrr|r} 1 & 0 & -1 & 0 \\ 0 & 1 & -\sqrt{2} & 0 \\ 0 & 0 & 0 & 0 \end{array}\right]$$

All solutions are given by $(s, s\sqrt{2}, s)$, where s is in **R**. A basis is obtained by setting $s = 1$: $\{(1, \sqrt{2}, 1)\}$

5. From the result of Problem 3 (or simply because it is symmetric), A is diagonalizable. By the procedure of Section 8.3, the desired matrices P and D are obtained from, respectively, the basis of eigenvectors of Problem 3 and the eigenvalues to which they correspond. The matrix P is the matrix of columns of those eigenvectors, and D is the diagonal matrix with the eigenvalues in the corresponding order:

$$P = \begin{bmatrix} -1 & -1 & 1 \\ 0 & 0 & 1 \\ 0 & 1 & 1 \end{bmatrix}, \quad D = \begin{bmatrix} 0 & 0 & 0 \\ 0 & 0 & 0 \\ 0 & 0 & 6 \end{bmatrix}$$

6. Since A is not symmetric, we do not know beforehand whether or not A is diagonalizable. We must first calculate its eigenvalues and see if a basis of eigenvectors exists. To find the eigenvalues, we solve $0 = \det(A - \lambda I)$ and then find a basis for each eigenspace.

$$0 = \det\begin{bmatrix} 1-\lambda & 0 & 0 \\ 0 & -\lambda & 1 \\ 0 & -1 & 2-\lambda \end{bmatrix} = (1-\lambda)^3$$

Thus, the only eigenvalue is $\lambda = 1$. Calculating a basis for its eigenspace, we have:

$$\lambda = 1: \quad \left[\begin{array}{ccc|c} 0 & 0 & 0 & 0 \\ 0 & -1 & 1 & 0 \\ 0 & -1 & 1 & 0 \end{array}\right] \rightarrow \left[\begin{array}{ccc|c} 0 & 1 & 1 & 0 \\ 0 & 0 & 0 & 0 \\ 0 & 0 & 0 & 0 \end{array}\right]$$

All solutions are given by $(s, -t, t)$, where s and t are in $\mathbf{R}$. A basis is obtained by setting $s = 1, t = 0$ and then $s = 0, t = 1$: $\{(1, 0, 0), (0, -1, 1)\}$. Since this is not a basis for $\mathbf{R}^3$, no basis of eigenvectors exists. Thus A is not diagonalizable.

7. **(a)** Yes. For each i and j, $b_{ij} = b_{ji}$.

(b) B is not orthogonal. For example, $\mathbf{b}^1 \cdot \mathbf{b}^3 = 1 \neq 0$.

8. We follow the procedure of Section 8.5. First, we apply the Gram-Schmidt process to the columns of B, which gives the following *orthogonal* basis for its column space: $\{(0, 0, 1), (0, 1, 0), (1, 0, 0)\}$. Because this is also an *orthonormal* basis, its vectors become the columns of Q:

$$Q = \begin{bmatrix} 0 & 0 & 1 \\ 0 & 1 & 0 \\ 1 & 0 & 0 \end{bmatrix}.$$

The upper-triangular matrix R has the following entries

$$r_{11} = \mathbf{b}^1 \cdot \mathbf{q}^1 = (0, 0, 1) \cdot (0, 0, 1) = 1$$
$$r_{12} = \mathbf{b}^2 \cdot \mathbf{q}^1 = (0, 1, 0) \cdot (0, 0, 1) = 0$$
$$r_{13} = \mathbf{b}^3 \cdot \mathbf{q}^1 = (1, 0, 1) \cdot (0, 0, 1) = 1$$

$$r_{22} = \mathbf{b}^2 \cdot \mathbf{q}^2 = (0, 1, 0) \cdot (0, 1, 0) = 1$$
$$r_{23} = \mathbf{b}^3 \cdot \mathbf{q}^2 = (1, 0, 1) \cdot (0, 1, 0) = 0$$
$$r_{33} = \mathbf{b}^3 \cdot \mathbf{q}^3 = (1, 0, 1) \cdot (1, 0, 0) = 1$$

and all other entries of R are equal to 0. Thus,

$$R = \begin{bmatrix} 1 & 0 & 1 \\ 0 & 1 & 0 \\ 0 & 0 & 1 \end{bmatrix}$$

9. For the given matrix B, we have B = QR, where Q and R are the matrices found in solving Problem 8. Thus, the first iterate of the QR algorithm is given by

$$RQ = \begin{bmatrix} 1 & 0 & 1 \\ 0 & 1 & 0 \\ 0 & 0 & 1 \end{bmatrix}\begin{bmatrix} 0 & 0 & 1 \\ 0 & 1 & 0 \\ 1 & 0 & 0 \end{bmatrix} = \begin{bmatrix} 1 & 0 & 1 \\ 0 & 1 & 0 \\ 1 & 0 & 0 \end{bmatrix}$$

10. **(a)** $\mathbf{w} \cdot \mathbf{z} = (i, 1, 1 + i) \cdot (-i, i, -i)$

$$= (i)(i) + (1)(-i) + (1 + i)(i) = -2$$

(b) $\|\mathbf{w}\| = \sqrt{(i, 1, 1 + i) \cdot (i, 1, 1 + i)} = \sqrt{(i)(-i) + (1)(1) + (1 + i)(1 - i)}$

$$= \sqrt{1 + 1 + 2} = 2$$

(c) $A^2 = \begin{bmatrix} i & -i \\ i & i \end{bmatrix}\begin{bmatrix} i & -i \\ i & i \end{bmatrix} = \begin{bmatrix} (i)(i) + (-i)(i) & (i)(-i) + (-i)(i) \\ (i)(i) + (i)(i) & (i)(-i) + (i)(i) \end{bmatrix} = \begin{bmatrix} 0 & 2 \\ -2 & 0 \end{bmatrix}$

(d) To find A^{-1}, we form $[A|I]$ and transform it to row-reduced echelon form:

$$\left[\begin{array}{cc|cc} i & -i & 1 & 0 \\ i & i & 0 & 1 \end{array}\right] \rightarrow \left[\begin{array}{cc|cc} 1 & 0 & -i/2 & -i/2 \\ 0 & 1 & i/2 & -i/2 \end{array}\right]$$

Thus, $A^{-1} = \begin{bmatrix} -i/2 & -i/2 \\ i/2 & -i/2 \end{bmatrix}$

(e) A is not hermitian; the diagonal entries are not all real, so A cannot be equal to A^H.

(f) A is not unitary: $\mathbf{a}^1 \cdot \mathbf{a}^2 = 0$, so the columns are orthogonal, but they are not of length 1.

11. Solve $\det(B - \lambda I) = 0$:

$$\det \begin{bmatrix} i-\lambda & 0 & 0 \\ 0 & -\lambda & i \\ 0 & i & -\lambda \end{bmatrix} = (i - \lambda)(\lambda^2 + 1) = 0$$

Thus, the eigenvalues are ±i.

$$\lambda = i: \quad \left[\begin{array}{ccc|c} 0 & 0 & 0 & 0 \\ 0 & -i & i & 0 \\ 0 & i & -i & 0 \end{array}\right] \rightarrow \left[\begin{array}{ccc|c} 0 & 1 & -1 & 0 \\ 0 & 0 & 0 & 0 \\ 0 & 0 & 0 & 0 \end{array}\right]$$

All solutions are given by (s, t, t), for s and t in **C**. A basis is obtained by setting $s = 1, t = 0$ and $s = 0, t = 1$: $\{(1, 0, 0), (0, 1, 1)\}$

$$\lambda = -i: \quad \left[\begin{array}{ccc|c} 2i & 0 & 0 & 0 \\ 0 & i & i & 0 \\ 0 & i & i & 0 \end{array}\right] \rightarrow \left[\begin{array}{ccc|c} 1 & 0 & 0 & 0 \\ 0 & 1 & 1 & 0 \\ 0 & 0 & 0 & 0 \end{array}\right]$$

All solutions are given by $(0, -s, s)$, for s in **C**. A basis is obtained by setting $s = 1$: $\{(0, -1, 1)\}$.

12. From Problem 11, there is a basis for $\mathbf{C}^3$ that consists of eigenvectors of B, so B is diagonalizable. By the diagonalization procedure of Section 8.3 extended to the complex numbers in Section 8.6, the matrix of symmetry P can be given by the columns of the basis vectors of Problem 11. The diagonalization matrix D then has the eigenvalues as its diagonal entries in corresponding order.

$$P = \begin{bmatrix} 1 & 0 & 0 \\ 0 & 1 & -1 \\ 0 & 1 & 1 \end{bmatrix}, \quad D = \begin{bmatrix} i & 0 & 0 \\ 0 & i & 0 \\ 0 & 0 & -i \end{bmatrix}$$

Answers to Odd-Numbered Review Exercises

1. For the characteristic polynomial we have

$$\det(A - \lambda I) = \det\begin{bmatrix} 2-\lambda & 1 \\ 1 & 2-\lambda \end{bmatrix}$$

The characteristic equation is obtained by setting this polynomial equal to 0, and its solutions are the eigenvalues:

$$0 = \lambda^2 - 4\lambda + 3 = (\lambda - 1)(\lambda - 3)$$

Therefore, the eigenvalues are $\lambda = 1$ and $\lambda = 3$.

3. For the characteristic equation

$$0 = \det(A - \lambda I) = \det\begin{bmatrix} 1-\lambda & 0 & 0 \\ -1 & 2-\lambda & -1 \\ 0 & 0 & 1-\lambda \end{bmatrix} = (2 - \lambda)(1 - \lambda)^2,$$

the eigenvalues are $\lambda = 1$ and $\lambda = 2$.

5. Since this matrix is upper triangular, the characteristic equation is simply

$$0 = (1 - \lambda)(3 - \lambda)(5 - \lambda)(7 - \lambda)$$

and the eigenvalues are $\lambda = 1, 3, 5$, and 7.

7. For each eigenvalue λ, we find a basis for the solution space of the equation $(A - \lambda I)\mathbf{x} = \mathbf{0}$ by transforming $[A - \lambda I \mid \mathbf{0}]$ into row-reduced echelon form.

$$\lambda = 1: \quad \left[\begin{array}{cc|c} 1 & 1 & 0 \\ 1 & 1 & 0 \end{array}\right] \rightarrow \left[\begin{array}{cc|c} 1 & 1 & 0 \\ 0 & 0 & 0 \end{array}\right]$$

Interpreting this as a system of linear equations in x and y, we have simply $x + y = 0$. Thus, $y = -x$, and the solutions have the form $(-s, s)$ for all real numbers s. A basis for this eigenspace can be obtained by setting $s = 1$: $\{(-1, 1)\}$

$$\lambda = 3: \left[\begin{array}{rr|r} -1 & 1 & 0 \\ 1 & -1 & 0 \end{array}\right] \rightarrow \left[\begin{array}{rr|r} 1 & -1 & 0 \\ 0 & 0 & 0 \end{array}\right]$$

Proceeding as above, we obtain solutions of the form (s, s) for s in **R**, and the basis for the eigenspace: $\{(1, 1)\}$.

9. Each matrix in Exercises 1, 2, 4, 5, and 6 has a complete set of *distinct* eigenvalues. Therefore, each is diagonalizable over the real (and complex) numbers. For Exercise 3, we note that there are two vectors in a basis for the eigenspace corresponding to $\lambda = 1$ and one more corresponding to the other eigenvalue, $\lambda = 2$. Thus, there is a basis of eigenvectors for $\mathbf{R}^3$, and this matrix is also diagonalizable (over both real and complex numbers).

11. For each eigenvalue ($\lambda = 0, 1, 3$), we find a nonzero vector in the solution space of $(A - \lambda I)\mathbf{x} = \mathbf{0}$.

$$\lambda = 0: \left[\begin{array}{rrr|r} 2 & 1 & 1 & 0 \\ 1 & 1 & 0 & 0 \\ 1 & 0 & 1 & 0 \end{array}\right] \rightarrow \left[\begin{array}{rrr|r} 1 & 0 & 1 & 0 \\ 0 & 1 & -1 & 0 \\ 0 & 0 & 0 & 0 \end{array}\right] \rightarrow (-1, 1, 1)$$

$$\lambda = 1: \left[\begin{array}{rrr|r} 1 & 1 & 1 & 0 \\ 1 & 0 & 0 & 0 \\ 1 & 0 & 0 & 0 \end{array}\right] \rightarrow \left[\begin{array}{rrr|r} 1 & 0 & 0 & 0 \\ 0 & 1 & 1 & 0 \\ 0 & 0 & 0 & 0 \end{array}\right] \rightarrow (0, -1, 1)$$

$$\lambda = 3: \left[\begin{array}{rrr|r} -1 & 1 & 1 & 0 \\ 1 & -2 & 0 & 0 \\ 1 & 0 & -2 & 0 \end{array}\right] \rightarrow \left[\begin{array}{rrr|r} 1 & 0 & -2 & 0 \\ 0 & 1 & -1 & 0 \\ 0 & 0 & 0 & 0 \end{array}\right] \rightarrow (2, 1, 1)$$

The desired matrix consists of columns of these vectors:

$$P = \begin{bmatrix} -1 & 0 & 2 \\ 1 & -1 & 1 \\ 1 & 1 & 1 \end{bmatrix}$$

13. Yes. Letting $\mathbf{a}^1$, $\mathbf{a}^2$, and $\mathbf{a}^3$ be the columns of the matrix, by Theorem 5 we need only verify that:

$$\mathbf{a}^1 \cdot \mathbf{a}^2 = \mathbf{a}^2 \cdot \mathbf{a}^3 = \mathbf{a}^1 \cdot \mathbf{a}^3 = 0$$

and $\quad \|\mathbf{a}^1\| = \|\mathbf{a}^2\| = \|\mathbf{a}^3\| = 1$

These six computations are easily checked to conclude the result.

15. The two orthogonally diagonalizable matrices identified in Exercise 12 are from Exercises 1 and 6. An orthogonal matrix for each can be obtained by normalizing the columns of the matrix of eigenvectors which correspond to the eigenvalues. These eigenvectors were calculated in Exercises 7 and 11, respectively.

Thus, for the matrix of Exercise 1,

$$\begin{bmatrix} -1 & 1 \\ 1 & 1 \end{bmatrix} \rightarrow \begin{bmatrix} -1/\sqrt{2} & 1/\sqrt{2} \\ 1/\sqrt{2} & 1/\sqrt{2} \end{bmatrix}$$

and for the matrix of Exercise 6,

$$\begin{bmatrix} -1 & 0 & 2 \\ 1 & -1 & 1 \\ 1 & 1 & 1 \end{bmatrix} \rightarrow \begin{bmatrix} -1/\sqrt{3} & 0 & 2/\sqrt{6} \\ 1/\sqrt{3} & -1/\sqrt{2} & 1/\sqrt{6} \\ 1/\sqrt{3} & 1/\sqrt{2} & 1/\sqrt{6} \end{bmatrix}$$

17. **(a)** Following the procedure of Section 8.5, we apply the Gram-Schmidt process to the columns of the matrix A of Exercise 1, (2, 1) and (1, 2), to obtain the following *orthogonal* basis for its column space: {(2, 1), (1, -2)}. Normalizing this set (dividing each vector by its length), we obtain the *orthonormal* basis, $\{(2/\sqrt{5}, 1/\sqrt{5}), (1/\sqrt{5}, -2/\sqrt{5})\}$, which forms the columns of Q:

$$Q = \begin{bmatrix} 2/\sqrt{5} & 1/\sqrt{5} \\ 1/\sqrt{5} & -2/\sqrt{5} \end{bmatrix}$$

The upper-triangular matrix R has the following entries

$$r_{11} = \mathbf{a}^1 \cdot \mathbf{q}^1 = (2, 1) \cdot (2/\sqrt{5}, 1/\sqrt{5}) = 5/\sqrt{5}$$
$$r_{12} = \mathbf{a}^2 \cdot \mathbf{q}^1 = (1, 2) \cdot (2/\sqrt{5}, 1/\sqrt{5}) = 4/\sqrt{5}$$
$$r_{22} = \mathbf{a}^2 \cdot \mathbf{q}^2 = (1, 2) \cdot (1/\sqrt{5}, -2/\sqrt{5}) = -3/\sqrt{5}$$

and $r_{21} = 0$. Thus,

$$R = \begin{bmatrix} 5/\sqrt{5} & 4/\sqrt{5} \\ 0 & -3/\sqrt{5} \end{bmatrix}$$

(b) For the given matrix A, we have A = QR, where Q and R are the matrices found in part (a). Thus, the first iterate of the QR algorithm is given by

$$RQ = \begin{bmatrix} 5/\sqrt{5} & 4/\sqrt{5} \\ 0 & -3/\sqrt{5} \end{bmatrix} \begin{bmatrix} 2/\sqrt{5} & 1/\sqrt{5} \\ 1/\sqrt{5} & -2/\sqrt{5} \end{bmatrix} = \begin{bmatrix} 14/5 & -3/5 \\ -3/5 & 6/5 \end{bmatrix}$$

19. **(a)** Following the procedure of Section 8.5, we apply the Gram-Schmidt process to the columns of the given matrix B of Exercise 3, and obtain the following *orthogonal* basis for its column space: {(1, -1, 0), (1, 1, 0), (0, 0, -1)}. Normalizing each vector (by dividing it by its length) gives the *orthonormal* basis, $\{(1/\sqrt{2}, -1/\sqrt{2}, 0), (1/\sqrt{2}, 1/\sqrt{2}, 0), (0, 0, -1)\}$, which forms the columns of Q:

$$Q = \begin{bmatrix} 1/\sqrt{2} & 1/\sqrt{2} & 0 \\ -1/\sqrt{2} & 1/\sqrt{2} & 0 \\ 0 & 0 & -1 \end{bmatrix}$$

The upper-triangular matrix R has the following entries

$$\begin{aligned}
r_{11} &= \mathbf{b}^1 \cdot \mathbf{q}^1 = (1, -1, 0) \cdot (1/\sqrt{2}, -1/\sqrt{2}, 0) = 2/\sqrt{2} \\
r_{12} &= \mathbf{b}^2 \cdot \mathbf{q}^1 = (0, 2, 0) \cdot (1/\sqrt{2}, -1/\sqrt{2}, 0) = -2/\sqrt{2} \\
r_{13} &= \mathbf{b}^3 \cdot \mathbf{q}^1 = (0, 1, 1) \cdot (1/\sqrt{2}, -1/\sqrt{2}, 0) = -1/\sqrt{2} \\
r_{22} &= \mathbf{b}^2 \cdot \mathbf{q}^2 = (0, 2, 0) \cdot (1/\sqrt{2}, 1/\sqrt{2}, 0) = 2/\sqrt{2} \\
r_{23} &= \mathbf{b}^3 \cdot \mathbf{q}^2 = (0, 1, 1) \cdot (1/\sqrt{2}, 1/\sqrt{2}, 0) = 1/\sqrt{2} \\
r_{33} &= \mathbf{b}^3 \cdot \mathbf{q}^3 = (0, 1, 1) \cdot (0, 0, -1) = -1
\end{aligned}$$

and all other entries of R are equal to 0. Thus,

$$R = \begin{bmatrix} \sqrt{2} & -\sqrt{2} & -\sqrt{2}/2 \\ 0 & \sqrt{2} & \sqrt{2}/2 \\ 0 & 0 & -1 \end{bmatrix}$$

(b) For the given matrix B, we have B = QR, where Q and R are the matrices found in part (a). Thus, the first iterate of the QR algorithm is given by

$$\mathrm{RQ} = \begin{bmatrix} \sqrt{2} & -\sqrt{2} & -\sqrt{2}/2 \\ 0 & \sqrt{2} & \sqrt{2}/2 \\ 0 & 0 & -1 \end{bmatrix} \begin{bmatrix} 1/\sqrt{2} & 1/\sqrt{2} & 0 \\ -1/\sqrt{2} & 1/\sqrt{2} & 0 \\ 0 & 0 & -1 \end{bmatrix} = \begin{bmatrix} 2 & 0 & \sqrt{2}/2 \\ -1 & 1 & -\sqrt{2}/2 \\ 0 & 0 & 1 \end{bmatrix}$$

21. This set is a basis. One way to verify this is to check that the determinant of the matrix of columns is nonzero:

$$\det \begin{bmatrix} 1 & 1 & 1 \\ i & 1+i & 1-i \\ -i & 1-i & 1+i \end{bmatrix} = \det \begin{bmatrix} 1 & 0 & 0 \\ i & 1 & 1-2i \\ -i & 0 & 1+2i \end{bmatrix}$$

$$= \det \begin{bmatrix} 1 & 1-2i \\ 1 & 1+2i \end{bmatrix} = 4i \neq 0$$

23. The linear transformation is not one-to one. One way to verify this is to check that $\det \mathrm{A} = 0$, which tells us that $\mathrm{A}\mathbf{x} = \mathbf{0}$ has more than one solution:

$$\det \begin{bmatrix} i & -i & 1 \\ i & 1+i & 1-i \\ 0 & 2+i & 1 \end{bmatrix} = \det \begin{bmatrix} i & -i & 0 \\ 0 & 2+i & 1 \\ 0 & 2+i & 1 \end{bmatrix} = 0$$

(because the second and third rows are equal).

25. To find the eigenvalues of the given matrix A, we solve $\det(\mathrm{A} - \lambda\mathrm{I}) = 0$; that is,

$$\det \begin{bmatrix} 1-\lambda & 0 & i \\ 0 & 1-\lambda & 0 \\ -i & 0 & 2-\lambda \end{bmatrix} = (1 - \lambda)(\lambda^2 - 3\lambda + 1) = 0$$

Thus, the eigenvalues are $\lambda = 1, (3 \pm \sqrt{5})/2$. For each eigenvalue, we find the corresponding eigenspace (all solutions to $(\mathrm{A} - \lambda\mathrm{I})\mathbf{x} = \mathbf{0}$) and a basis for that eigenspace.

$\lambda_1 = 1;\quad \mathbf{x}_1 = (0, c, 0),\ c$ in $\mathbf{C}$; Basis: $\{(0, 1, 0)\}$

$\lambda_2 = (3 + \sqrt{5})/2;\quad \mathbf{x}_2 = (c(-1 + \sqrt{5})i/2, 0, c),$ c in $\mathbf{C}$;
Basis: $\{((-1 + \sqrt{5})i/2, 0, 1)\}$

$\lambda_3 = (3 - \sqrt{5})/2;\quad \mathbf{x}_3 = (c(-1 - \sqrt{5})i/2, 0, c),\ c$ in $\mathbf{C}$;
Basis: $\{((-1 - \sqrt{5})i/2, 0, 1)\}$

27. In each case, there are three distinct eigenvalues for the 3×3 matrix. By Theorem 3 of Section 8.6, each is diagonalizable.

29. The given hermitian matrix is that of Exercise 25. The desired unitary matrix P can be obtained from the resulting basis vectors of that exercise by normalizing each (dividing it by its length) and forming the matrix of columns of the normalized vectors:

$$\begin{bmatrix} 0 & \frac{(-1+\sqrt{5})i}{2} & \frac{(-1-\sqrt{5})i}{2} \\ 1 & 0 & 0 \\ 0 & 1 & 1 \end{bmatrix} \rightarrow \begin{bmatrix} 0 & \frac{\alpha(-1+\sqrt{5})i}{2} & \frac{\beta(-1-\sqrt{5})i}{2} \\ 1 & 0 & 0 \\ 0 & \alpha & \beta \end{bmatrix}$$

where $\alpha = \dfrac{\sqrt{2}}{\sqrt{5-\sqrt{5}}}$ and $\beta = \dfrac{\sqrt{2}}{\sqrt{5+\sqrt{5}}}$

Answers to Even-Numbered Exercises

SECTION 8.1

2. Eigenvalue: 0; eigenvectors: $\mathbf{x} \neq \mathbf{0}$ in $\mathbf{R}^n$.

4. Eigenvalue: -1; eigenvectors: $\mathbf{x} \neq \mathbf{0}$ in $\mathbf{R}^n$.

6. Eigenvalue: 1; eigenvectors: $\mathbf{x} = (c, c), c \neq 0$.
Eigenvalue: -1; eigenvectors: $\mathbf{x} = (c, -c), c \neq 0$.

8. Eigenvalue: 0; eigenvectors: $X = \begin{bmatrix} 0 & 0 \\ 0 & c \end{bmatrix},\ c \neq 0$

Eigenvalue: 1; eigenvectors: $X = \begin{bmatrix} c & 0 \\ 0 & c \end{bmatrix}$, $c \neq 0$

Eigenvalue: -1; eigenvectors: $X = \begin{bmatrix} 0 & c \\ c & c \end{bmatrix}$, $c \neq 0$

10. $\lambda^2 + 3 = 0$; no real eigenvalues

12. $\lambda^2 = 0$; $\lambda = 0$

14. $-\lambda^3 + 6\lambda^2 - 10\lambda + 4 = 0$; $\lambda = 2, 2 \pm \sqrt{2}$

16. $-\lambda^3 + 3\lambda^2 - 3\lambda = 0$; $\lambda = 0$

18. $(1 - \lambda)(2 - \lambda)^2(-1 - \lambda) = 0$; $\lambda = \pm 1, 2$

20. $(1 - \lambda)^2(\lambda^2 + \lambda + 1) = 0$; $\lambda = 1$

22. No eigenvectors because there are no eigenvalues.

24. $\lambda = 0$: all (c, d), c, d in **R**, but not both 0; basis: $\{(1, 0), (0, 1)\}$

26. $\lambda = 2$: all $(-c, 0, c)$, $c \neq 0$ in **R**; basis: $\{(-1, 0, 1)\}$
$\lambda = 2 + \sqrt{2}$: all $(c, -c\sqrt{2}, c)$, $c \neq 0$ in **R**; basis: $\{(1, -\sqrt{2}, 1)\}$
$\lambda = 2 - \sqrt{2}$: all $(c, c\sqrt{2}, c)$, $c \neq 0$ in **R**; basis: $\{(1, \sqrt{2}, 1)\}$

28. $\lambda = 0$: all $(c, -3c, c)$, $c \neq 0$ in **R**; basis: $\{(1, -3, 1)\}$

30. $\lambda = 1$: all $(2c, 0, 3c, 4c)$, $c \neq 0$ in **R**; basis: $\{(2, 0, 3, 4)\}$
$\lambda = -1$: all $(0, 0, c, 0)$, $c \neq 0$ in **R**; basis: $\{(0, 0, 1, 0)\}$
$\lambda = 2$: all $(0, 3c, c, d)$, c, d in **R**, but not both 0; basis: $\{(0, 3, 1, 0), (0, 0, 0, 1)\}$

32. $\lambda = 1$: all (c, d, d, d), c, d in **R**, but not both 0; basis: $\{(1, 0, 0, 0), (0, 1, 1, 1)\}$

34. 1, 5, -6

36. -1, 0, 2

40. The eigenvalues are $\lambda = \pm 1$. The eigenspace for $\lambda = 1$ is the line that forms an angle θ with the positive x-axis. The eigenspace for $\lambda = -1$ is the line that forms an angle $\pi/2 + \theta$ with the positive x-axis.

SECTION 8.2

2. $t_{23} = 0.1$

4. $T^2 = \begin{bmatrix} .15 & .12 & .16 \\ .62 & .73 & .58 \\ .23 & .15 & .26 \end{bmatrix}$

6. (0, 1/2, 1/2)

8. (3/20, 3/5, 1/4)

10. 0.199

12. 1/2; 0

14. 7/18

16. (0, 0, 1)

18. (9/31, 10/31, 12/31)

SECTION 8.3

2. Diagonalizable;

$$P = \begin{bmatrix} 1 & 1 \\ -1 & 1 \end{bmatrix}, \quad D = \begin{bmatrix} -1 & 0 \\ 0 & 1 \end{bmatrix}$$

4. Not diagonalizable.

6. Diagonalizable;

$$P = \begin{bmatrix} 1 & -1 & 1 \\ \sqrt{2} & 0 & -\sqrt{2} \\ 1 & 1 & 1 \end{bmatrix}, \quad D = \begin{bmatrix} 2-\sqrt{2} & 0 & 0 \\ 0 & 0 & 0 \\ 0 & 0 & 2+\sqrt{2} \end{bmatrix}$$

8. Diagonalizable;

$$P = \begin{bmatrix} -1 & 1 & 1 \\ 1 & 0 & -1 \\ 0 & 1 & 2 \end{bmatrix}, \quad D = \begin{bmatrix} 0 & 0 & 0 \\ 0 & 0 & 0 \\ 0 & 0 & -2 \end{bmatrix}$$

10. Not diagonalizable.

12. Diagonalizable;

$$P = \begin{bmatrix} 0 & 1 & 0 & 0 \\ 1 & 0 & 3 & 0 \\ 1 & 0 & 1 & 0 \\ 0 & 2 & 0 & 1 \end{bmatrix}, \quad D = \begin{bmatrix} -1 & 0 & 0 & 0 \\ 0 & 1 & 0 & 0 \\ 0 & 0 & 2 & 0 \\ 0 & 0 & 0 & 2 \end{bmatrix}$$

14. Orthogonal

16. Not orthogonal

18. $\begin{bmatrix} 1/\sqrt{2} & 1/\sqrt{2} \\ -1/\sqrt{2} & 1/\sqrt{2} \end{bmatrix}$

20. $\begin{bmatrix} 1/2 & -1/\sqrt{2} & 1/2 \\ 1/\sqrt{2} & 0 & -1/\sqrt{2} \\ 1/2 & 1/\sqrt{2} & 1/2 \end{bmatrix}$

22. $\begin{bmatrix} 1/\sqrt{2} & 1/\sqrt{2} & 0 \\ -1/\sqrt{2} & 1/\sqrt{2} & 0 \\ 0 & 0 & 1 \end{bmatrix}$

24. $\begin{bmatrix} 1/\sqrt{2} & 1/\sqrt{6} & 1/\sqrt{12} & 1/2 \\ -1/\sqrt{2} & 1/\sqrt{6} & 1/\sqrt{12} & 1/2 \\ 0 & -2/\sqrt{6} & 1/\sqrt{12} & 1/2 \\ 0 & 0 & -3/\sqrt{12} & 1/2 \end{bmatrix}$

SECTION 8.4

2. $\dfrac{(x')^2}{12} + \dfrac{(y')^2}{22} + \dfrac{(z')^2}{132} = 1$ (Ellipsoid)

4. $\dfrac{(x')^2}{2} - \dfrac{(y')^2}{4} = 1$ (Hyperbolic cylinder)

6. $\dfrac{(x')^2}{2} - \dfrac{(y')^2}{2} - \dfrac{(z')^2}{4} = 1$ (Hyperboloid of two sheets)

8. $-\sqrt{3}(x'')^2 = y''$ (Parabolic cylinder)

10. $\dfrac{4(x'')^2}{43} + \dfrac{2(y'')^2}{43} + \dfrac{(z'')^2}{43} = 1$ (Ellipsoid)

12. $\mathbf{x}^T A\mathbf{x} + B\mathbf{x} = f$, where

$f = 1, \quad B = [5 \quad -2],$

and

$$A = \begin{bmatrix} 3 & 2 \\ 2 & -1 \end{bmatrix}.$$

16. $(x')^2 - 2x' + 4y' = 3$

18. $-(1/4)(x'')^2 = y''$ (Parabola)

SECTION 8.5

2. $Q = \begin{bmatrix} 3/5 & 4/5 \\ 4/5 & -3/5 \end{bmatrix}, \quad R = \begin{bmatrix} 5 & 2 \\ 0 & 1 \end{bmatrix}$

4. $Q = \begin{bmatrix} 1/\sqrt{2} & 1/\sqrt{2} & 0 \\ -1/\sqrt{2} & 1/\sqrt{2} & 0 \\ 0 & 0 & -1 \end{bmatrix}, \quad R = \begin{bmatrix} \sqrt{2} & -\sqrt{2} & -\sqrt{2}/2 \\ 0 & \sqrt{2} & \sqrt{2}/2 \\ 0 & 0 & -1 \end{bmatrix}$

6. To two decimal places:

$$Q = \begin{bmatrix} 0.86 & -0.50 & -0.12 \\ 0.43 & 0.57 & 0.70 \\ 0.29 & 0.65 & -0.70 \end{bmatrix}, \quad R = \begin{bmatrix} 1.17 & 0.64 & 0.45 \\ 0 & 0.10 & 0.11 \\ 0 & 0 & 0.00 \end{bmatrix}$$

8. $\begin{bmatrix} 23/5 & 14/5 \\ 4/5 & -3/5 \end{bmatrix}$

10. $\begin{bmatrix} 2 & 0 & \sqrt{2}/2 \\ -1 & 1 & -\sqrt{2}/2 \\ 0 & 0 & 1 \end{bmatrix}$

12. To two decimal places:

$$\begin{bmatrix} 1.40 & 0.07 & 0.00 \\ 0.07 & 0.13 & 0.00 \\ 0.00 & 0.00 & 0.00 \end{bmatrix}$$

14. $Q = \begin{bmatrix} 1 & 0 \\ 0 & 1 \end{bmatrix}, \quad R = \begin{bmatrix} 1 & 1 \\ 0 & 0 \end{bmatrix}$

SECTION 8.6

2. -2

4. 4

6. $(2/3 - i, -1/3, -(4/3)i)$

8. $\begin{bmatrix} i & 1-i & 1 \\ 2i & -1 & 1+2i \\ -1 & 1 & 2i \end{bmatrix}$

10. $\frac{1}{3}\begin{bmatrix} -2i & -i & 1 \\ i & -i & 1 \\ 1 & 1 & -2i \end{bmatrix}$

12. $-3i$

16. No

18. No

20. $\{(0, 1, 0, 0), (0, 0, 1, 0)\}$

22. $\{(i, 1, 1)\}$

24. $\lambda_1 = -2i$; $\mathbf{x}_1 = (s, is)$; basis: $\{(1, i)\}$
$\lambda_2 = 2i$; $\mathbf{x}_2 = (t, -it)$; basis: $\{(1, -i)\}$

26. $\lambda_1 = 1$; $\mathbf{x}_1 = (-r, 0, r)$; basis: $\{(-1, 0, 1)\}$
$\lambda_2 = 1 - \sqrt{2}$; $\mathbf{x}_2 = (s, (i\sqrt{2})s, s)$; basis: $\{(1, i\sqrt{2}, 1)\}$
$\lambda_3 = 1 + \sqrt{2}$; $\mathbf{x}_3 = (t, (-i\sqrt{2})t, t)$; basis: $\{(1, -i\sqrt{2}, 1)\}$

28. $\lambda_1 = 0$; $\mathbf{x}_1 = (0, 0, r)$; basis: $\{(0, 0, 1)\}$
$\lambda_2 = 1 + i$; $\mathbf{x}_2 = ((-2 + 2i)s, (-1 - i)s, s)$; basis: $\{(-2 + 2i, -1 - i, 1)\}$
$\lambda_3 = 1 - i$; $\mathbf{x}_3 = (0, (1 - i)t, t)$; basis: $\{(0, 1 - i, 1)\}$

30. $\lambda_1 = 1$; $\mathbf{x}_1 = (0, r, 0)$; basis: $\{(0, 1, 0)\}$
$\lambda_2 = -i$; $\mathbf{x}_2 = (-is, 0, s)$; basis: $\{(-i, 0, 1)\}$
$\lambda_3 = i$; $\mathbf{x}_3 = (it, 0, t)$; basis: $\{(i, 0, 1)\}$

32. $P = \begin{bmatrix} 1 & 1 \\ i & -i \end{bmatrix}$, $D = \begin{bmatrix} -2i & 0 \\ 0 & 2i \end{bmatrix}$

34. $P = \begin{bmatrix} -1/\sqrt{2} & 1/2 & 1/2 \\ 0 & i/\sqrt{2} & -i/\sqrt{2} \\ 1/\sqrt{2} & 1/2 & 1/2 \end{bmatrix}$, $D = \begin{bmatrix} 1 & 0 & 0 \\ 0 & 1-\sqrt{2} & 0 \\ 0 & 0 & 1+\sqrt{2} \end{bmatrix}$ (hermitian)

36. $P = \begin{bmatrix} -2+2i & 0 & 0 \\ -1-i & 1-i & 0 \\ 1 & 1 & 1 \end{bmatrix}$, $D = \begin{bmatrix} 1+i & 0 & 0 \\ 0 & 1-i & 0 \\ 0 & 0 & 0 \end{bmatrix}$

38. $P = \begin{bmatrix} 0 & -i & i \\ 1 & 0 & 0 \\ 0 & 1 & 1 \end{bmatrix}$, $D = \begin{bmatrix} 1 & 0 & 0 \\ 0 & -i & 0 \\ 0 & 0 & i \end{bmatrix}$

REVIEW EXERCISES

2. $\lambda^2 - 4 = 0;\quad \lambda = \pm 2$

4. $-(\lambda - 1)(\lambda - 2)(\lambda - 3) = 0;\quad \lambda = 1,\ \lambda = 2,\ \lambda = 3$

6. $-(\lambda - 1)\lambda(\lambda - 3) = 0;\quad \lambda = 0,\ \lambda = 1,\ \lambda = 3$

8. $\lambda = 1$: $\{(1, 1, 0), (1, 0, 1)\}$; $\lambda = 2$: $\{(0, 1, 0)\}$

10. $\begin{bmatrix} 1 & 1 & 0 \\ 1 & 0 & 1 \\ 0 & 1 & 0 \end{bmatrix}$

12. Exercises 1 and 6 only; the other matrices are not symmetric.

14. Not orthogonal

16. (a) $Q = \begin{bmatrix} 2/\sqrt{5} & 1/\sqrt{5} \\ 1/\sqrt{5} & -2/\sqrt{5} \end{bmatrix}$, $R = \begin{bmatrix} 5/\sqrt{5} & 4/\sqrt{5} \\ 0 & -3/\sqrt{5} \end{bmatrix}$

(b) $\begin{bmatrix} 2 & 1 \\ 1/5 & 2 \end{bmatrix}$

18. (a) $Q = \begin{bmatrix} 1/\sqrt{2} & 1/\sqrt{2} & 0 \\ -1/\sqrt{2} & 1/\sqrt{2} & 0 \\ 0 & 0 & -1 \end{bmatrix}$, $R = \begin{bmatrix} \sqrt{2} & -\sqrt{2} & -\sqrt{2}/2 \\ 0 & \sqrt{2} & \sqrt{2}/2 \\ 0 & 0 & -1 \end{bmatrix}$

(b) $\begin{bmatrix} 2 & 0 & \sqrt{2}/2 \\ -1 & 1 & -\sqrt{2}/2 \\ 0 & 0 & 1 \end{bmatrix}$

20. Is a basis

22. All $c(-2 + 6i, -2 + i, 5)$, c in $\mathbf{C}$

24. No. $\text{Dim}(\text{Im}(T)) = 3 - 1 = 2 < 3$

26. $\lambda_1 = 1$; $\mathbf{x}_1 = (c, 0, 0)$, c in $\mathbf{C}$; basis: $\{(1, 0, 0)\}$

$\lambda_2 = i$; $\mathbf{x}_2 = c(1 + i, -1, 0)$, c in $\mathbf{C}$; basis: $\{(1 + i, -1, 0)\}$

$\lambda_3 = 3$; $\mathbf{x}_3 = c(4 + 3i, 1 - 3i, 10)$, c in $\mathbf{C}$; basis: $\{(0, 1 - i, 1)\}$

28. Exercise 21; hermitian

CHAPTER 9

Linear Programming

Chapter Overview

The topics covered in this chapter are:

1. Basic linear programming terminology. [Section 9.1]
2. A linear programming problem in two variables; geometric interpretation. [9.1]
3. Statement of the general linear programming problem: optimize the *objective function* $z = \mathbf{c} \cdot \mathbf{x}$ subject to $A\mathbf{x} \leq \mathbf{b}$, $\mathbf{b} \geq \mathbf{0}$, $\mathbf{x} \geq \mathbf{0}$. [9.1]
4. The *simplex algorithm*; conditions for terminating the algorithm. [9.2]
5. Procedure for solving the two-phase problem. [9.3]
6. Proofs of theorems concerning successful termination of the simplex algorithm. [9.4]

Remarks

Since Dantzig first presented the simplex algorithm in 1947, it has become a workhorse for solving systems of linear inequalities, usually optimizing a linear function of the variables, that has gone far beyond the imaginations of the original researchers. Although variations are introduced to gain speed and/or eliminate accumulated round-off problems, the basic algorithm remains virtually intact. We present this algorithm in Section 9.2 after describing the setting in Section 9.1. Both the practical and the theoretical considerations of the simplex algorithm are surprisingly simple, which allows its study at some depth with only the background of Chapter 3. A negative aspect to this simplicity is that as an example of the usefulness of linear algebra, it is almost *too* simple.

Sections 9.1 and 9.2 must both be covered if this application is to be introduced at all. Section 9.3 then demonstrates how some other systems of inequalities can be converted into a form in which the same machinery applies. This section could be omitted, but it is not that difficult and does help show that the simplex technique has a wider application than the student may have thought. An instructor could go even further with this idea by showing how integer solution problems, so-called *integer programming problems*, can be included by appropriate extension of the basic linear programming concepts. (This is not covered in this text.)

The theory that justifies the simplex algorithm is presented in Section 9.4. This material consists of proofs of theorems that were stated in the first three sections, but it is not deeply theoretical. Understanding what the simplex indicators are measuring (the slope of the increase or decrease in the objective function along a particular line in $\mathbf{R}^n$) is really all there is to it. However, it may not be worth the effort. If time remains and the instructor is so motivated, it might be better to pursue the subject of optimizing linear functionals on convex sets from a more abstract standpoint (although this subject is not covered in this text).

Solutions to Self-Test

1. First, we plot the lines $x + 5y = 30$, $x + y = 6$, $5x + y = 20$, and then determine the half-planes (indicated by the arrows in Figure 1) corresponding to the given inequalities. The feasible region is the part of the first quadrant that is common to these three half-planes; it is shaded in Figure 1.

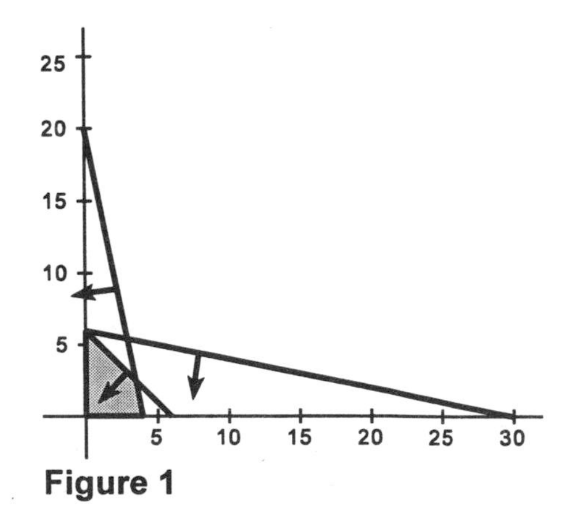

Figure 1

2. First, we plot the lines $x + 5y = 30$, $x + 2y = 6$, $-x + y = 4$, and then determine the half-planes (indicated by the arrows in Figure 2) corresponding to the given inequalities. The feasible region is empty because no part of the first quadrant is common to these three half-planes.

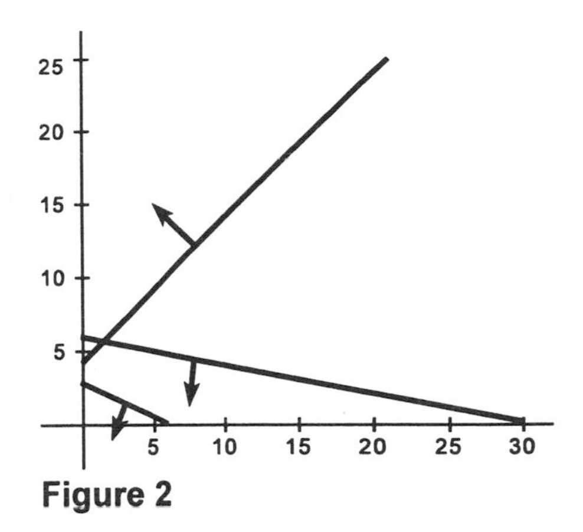

Figure 2

3. Solve the objective function for y, $y = -2x + z$, and superimpose this line on the region graphed in Exercise 1. Any such line has a slope of -2 and a y-intercept of z, the value of the objective function. Now, "slide" the line up or down to find the point or points in the feasible region for which the z value is the greatest. Notice that as we slide the line up, the y-intercept (z value) increases. The largest value occurs as the objective line is just about to leave the region, as it passes through the point (7/2, 5/2). The maximum value for z is obtained by substituting $x = 7/2$, $y = 5/2$ into the objective function, obtaining $z = 19/2$.

4. Here, solving the objective function for y yields $y = 2x - z$ and we need the point or points for which z is minimal. To obtain the smallest z, we slide the objective line until it is about to leave the region, when it "touches" the point (0, 5). The corresponding minimal value of z is -5.

5. We add a nonnegative slack variable to convert each inequality into an equation:

$$\begin{aligned} x + 5y + s_1 \qquad\qquad &= 30 \\ x + y \qquad + s_2 \qquad &= 6 \\ 5x + y \qquad\qquad + s_3 &= 20 \end{aligned}$$

6. From the x_2, x_3, and x_5 columns, we have $x_2 = 2$, $x_3 = 3$, $x_5 = 1$, and $x_1 = x_4 = x_6 = 0$. The value of the objective function is 10.

7. From the bottom row, we see that x_1, x_4, and x_6 will all increase the value of the objective function.

8. There are none; the current solution is minimal.

9. The current solution is minimal. There are no positive entries in the bottom row (except for the value of the objective function, 10).

10. The maximum rate of increase is 3, the rate for bringing in x_6. The new tableau, obtained from pivoting on the 3,6-entry, is as follows:

x_1	x_2	x_3	x_4	x_5	x_6	**b**
5/2	0	1	3/2	-1/2	0	5/2
-2	1	0	-2	-1	0	1
1/2	0	0	1/2	1/2	1	1/2
1/2	0	0	-1/2	3/2	0	23/2

Solutions to Odd-Numbered Review Exercises

1. First, we plot the lines $x + 10y = 30$, $x + y = 5$, $6x + y = 20$, and then determine the half-planes (indicated by the arrows in Figure 3) corresponding to the given inequalities. The feasible region is that part of the first quadrant which is common to these three half-planes; it is shaded in Figure 3.

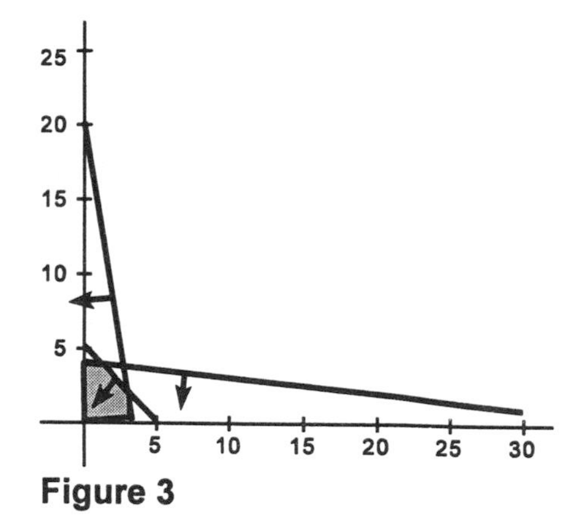

Figure 3

3. Since the objective function is $z = 2x + y$, we consider the family of parallel lines $y = -2x + z$, where z, the y-intercept for each line, is free to vary. The maximum and minimum values of the objective function are the largest and smallest values of this y-intercept for lines that intersect the region in at least one point. Since the y-intercept increases as we slide the line $y = -2x + z$ upward (as we can see from Figure 4), the smallest value occurs at the point (0, 0) and the largest at (3, 2). The corresponding minimum and maximum values can be obtained by substituting these ordered pairs into the objective function, which gives 0 and 8, respectively.

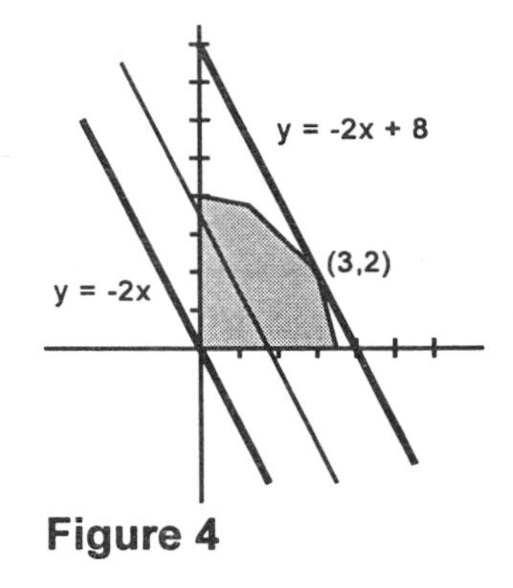

Figure 4

5. This time we slide the line $y = x - z$, which has slope 1 and y-intercept $-z$, up to make z larger and down to make it smaller. Notice that, as we move the line $y = x - z$ in an upward direction, it enters the region at the point (10/3, 0) and leaves the region at (0, 4). The first of these points gives a maximum value for z, namely 10/3; the second gives a minimum, -4.

7. We add a nonnegative slack variable to each inequality to convert it into an equation:

$$\begin{aligned} x + 10y + s_1 &= 40 \\ x + y + s_2 &= 5 \\ 6x + y + s_3 &= 20 \end{aligned}$$

9. The adjusted objective function vector is $\mathbf{c} = (2, 1, 0, 0, 0)$ and the vector of current costs is $\mathbf{c}^* = (0, 0, 0)$. The first tableau is the matrix of left-side coefficients A in the solution to Exercise 7, right side coefficients $\mathbf{b}$, and bottom row $\mathbf{c}^*\mathrm{A} - \mathbf{c}$ and $\mathbf{c}^*\mathbf{b}$:

x	y	s_1	s_2	s_3	**b**
1	10	1	0	0	40
1	1	0	1	0	5
6	1	0	0	1	20
-2	1	0	0	0	0

11. Starting with the tableau in Exercise 9, we notice that there are no positive simplex indicators (bottom row except for the last entry, which is the current

objective function value). Thus, the current solution, $x = y = 0$, with objective function value 0 is minimal.

The simplex indicator of maximum rate of increase is -2 in the first column, corresponding to variable x. Computing each ratio, 40/1, 5/1, 20/6, we see that the third row is one that will keep the solution feasible. That is, we pivot on the 3,1 position:

x	y	s_1	s_2	s_3	b
0	59/6	1	0	-1/6	110/3
0	5/6	0	1	-1/6	5/3
1	1/6	0	0	1/6	10/3
0	2/3	0	0	1/3	20/3

The only column that will increase the objective function value is the second, and the minimum of the ratios indicates that row 2 will keep the solution feasible, so we pivot on the 2,2 position:

x	y	s_1	s_2	s_3	b
0	0	1	-59/5	9/5	17
0	1	0	6/5	-1/5	2
1	0	0	-1/5	1/5	3
0	0	0	4/5	1/5	8

Since there are no negative simplex indicators, the solution is maximal. That is, $(x, y) = (3, 2)$ with the objective function value of 8.

13. We add slack variables to the first two constraints and subtract one from the third to convert to equations. Then, we could add an artificial variable w to the third constraint to obtain an embedded identity matrix and proceed with the two-phase method, using the first phase objective function $z = w$. Notice, however, that pivoting on the 3,2 position creates an embedded identity matrix in the second, third, and fourth columns. Thus, we need not carry out the first phase. The system of linear equations is as follows, where all variables are assumed to be nonnegative:

$$\begin{aligned} x + 10y + s_1 \qquad\qquad &= 40 \\ x + y \quad + s_2 \qquad &= 5 \\ -x + y \qquad\qquad - s_3 &= 1 \end{aligned}$$

Subtracting the third equation from the second and 10 times it from the first, we obtain an equivalent system with an embedded identity matrix (and nonnegative right side coefficients, so the solution remains feasible):

$$\begin{aligned} 11x \quad + s_1 \qquad + 10s_3 &= 30 \\ 2x \qquad + s_2 + s_3 &= 4 \\ -x + y \qquad\qquad - s_3 &= 1 \end{aligned}$$

The objective function remains the same, but with zero coefficients for the slack variables, so we prepare the first tableau with $\mathbf{c} = (1, -2, 0, 0, 0)$ and $\mathbf{c}^* = (0, 0, -1)$:

x	y	s_1	s_2	s_3	**b**
11	0	1	0	10	30
2	0	0	1	1	4
-1	1	0	0	-1	1
0	0	0	0	1	-1

There are no negative coefficients in the bottom row (except for the objective function value of -1), so the current solution is maximal. That is, $(x, y) = (0, 1)$ with value -1 is the (actually, *a*) maximal solution.

The 1 in the bottom row indicates that the minimum has not been achieved. So, we want to bring in the variable s_3, the fifth column. We choose the smaller of the nonnegative ratios 30/10 and 4/1 to decide on the pivot row. Thus, we pivot on the 10 in the fifth column, obtaining:

x	y	s_1	s_2	s_3	**b**
11/10	0	1/10	0	1	3
9/10	0	-1/10	1	0	1
1/10	1	1/10	0	0	4
-11/10	0	-1/10	0	0	-4

Since there are no positive coefficients in the bottom row, the solution $(x, y) = (0, 4)$ is minimal, with an objective function value of -4.

15. Here, only y is restricted to be nonnegative, so we replace x by the difference of two nonnegative variables, $x = x_1 - x_2$, and proceed as before. As in the solution to Exercise 13, we see that it is possible to avoid the two-phase method by "bringing in" y for the construction of the first tableau:

$$\begin{array}{rrrrrrl} 11x_1 & - 11x_2 & & + s_1 & & & = 30 \\ 2x_1 & - 2x_2 & & & + s_2 & + s_3 & = 4 \\ -x_1 & + x_2 & + y & & & - s_3 & = 1 \end{array}$$

Using the objective function $\mathbf{c} = (1, -1, -1, 0, 0, 0)$ and $\mathbf{c}^* = (0, 0, -1)$, we generate the first tableau:

x_1	x_2	y	s_1	s_2	s_3	**b**
11	-11	0	1	0	10	30
2	-2	0	0	1	1	4
-1	1	1	0	0	-1	1
0	0	0	0	0	1	-1

There are no negative simplex indicators in the bottom row, so the current solution, $x = x_1 - x_2 = 0 - 0 = 0$ and $y = 1$ is maximal and has value -1.

Decrease is indicated in the s_3 column and, to keep solutions feasible, we pivot on the first row (because $30/10 < 4/1$), sixth column. The new tableau is:

x_1	x_2	y	s_1	s_2	s_3	**b**
11/10	-11/10	0	1/10	0	1	3
9/10	-9/10	0	-1/10	1	0	1
1/10	-1/10	1	1/10	0	0	4
-11/10	11/10	0	-1/10	0	0	-4

The positive simplex indicator 11/10 that corresponds to x_2, together with its column of negative coefficients, imply that it is possible to move in that direction decreasing the objective function value in an unbounded manner.

Thus, no minimal value exists.

17. The first variable x_1 is unrestricted, so we replace it with the difference of two nonnegative variables, $x_1 = y_1 - y_2$. We subtract nonnegative slack variables from the first and third inequalities, and then add nonnegative artificial variables to each equation to obtain a representation with an embedded identity matrix. This situation is not equivalent to the original unless all artificial variables are 0, so we take as the objective function $z = w_1 + w_2 + w_3$ and apply the simplex algorithm to minimize it. In equation form, we have:

$$\begin{aligned}
y_1 - y_2 + 2x_2 - x_3 + 3x_4 - s_1 + a_1 &= 5 \\
y_1 - y_2 - 2x_2 + x_3 - x_4 + a_2 &= 10 \\
2y_1 - 2y_2 + x_2 + x_3 - x_4 - s_2 + a_3 &= 8
\end{aligned}$$

In tableau form, this becomes:

y_1	y_2	x_2	x_3	x_4	s_1	s_2	a_1	a_2	a_3	**b**
1	-1	2	-1	3	-1	0	1	0	0	5
1	-1	-2	1	-1	0	0	0	1	0	10
2	-2	1	1	-1	0	-1	0	0	1	8
4	-4	1	1	1	-1	-1	0	0	0	23

Pivoting on the 3,1 position of this tableau, then on the 1,5 position of the resulting one, then on the 1,7 position, and finally on the 2,4 position, we obtain the following tableau in which none of the artificial variables is involved:

y_1	y_2	x_2	x_3	x_4	s_1	s_2	a_1	a_2	a_3	**b**
0	0	-3	0	1	-1/2	1	1/2	3/2	-1	19/2
0	0	-2	1	-2	1/2	0	-1/2	1/2	0	5/2
1	-1	0	0	1	-1/2	0	1/2	1/2	0	15/2
0	0	0	0	0	0	0	-1	-1	-1	0

Thus, a feasible solution exists, and by discarding the columns of artificial variables and the bottom row, the starting tableau for a further optimization is available with the embedded identity matrix in the columns that correspond to y_1, x_3, and s_2.

Answers to Even-Numbered Exercises

SECTION 9.1

2. The feasible region is unbounded. See Figure 5.

4. There is no feasible region. See Figure 6.

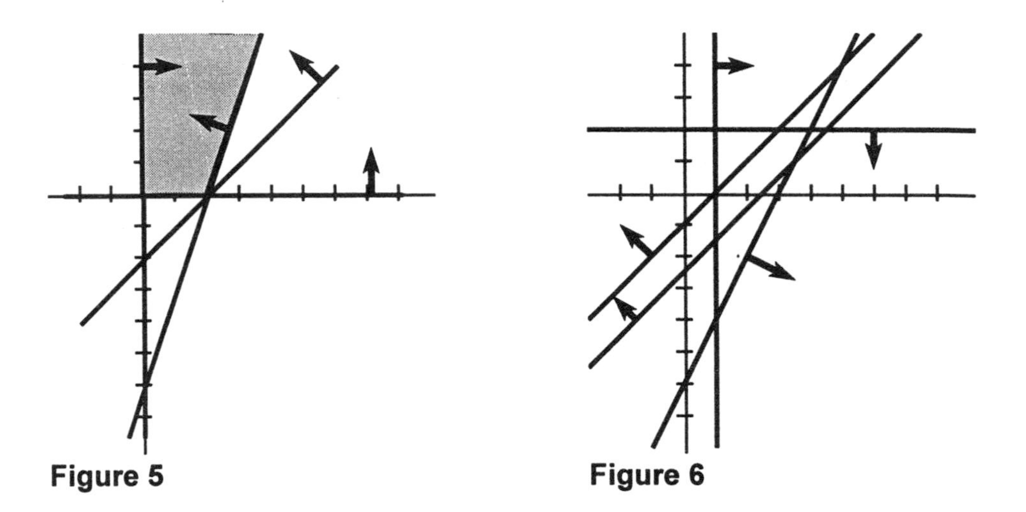

Figure 5 Figure 6

6. Since we want to maximize $z = 3x - y$, we consider the objective line $y = 3x - z$, which is parallel to the constraining line, $y = 3x - 6$. Notice that the value of z increases as the y-intercept of the objective line $(-z)$ decreases; that is, as the line $y = 3x - z$ moves to the right (and down). Just as this line is about to leave the feasible region, it coincides with the line $y = 3x - 6$, and consequently every point on the latter line gives the maximum value of z, namely 6.

8.
$$\begin{aligned} x - y + s_1 \quad &= 1 \\ x + 2y \quad + s_2 &= 8 \end{aligned}$$

$$x,\ y,\ s_1,\ s_2 \geq 0$$

10.
$$\begin{aligned} x - y + s_1 \quad\quad &= 2 \\ -x + 3y \quad + s_2 \quad &= 9 \\ 2x - y \quad\quad - s_3 &= 6 \end{aligned}$$

$$x,\ y,\ s_1,\ s_2,\ s_3 \geq 0$$

12. Row-reduced columns 1, 2, 3; $x_1 = 1,\ x_2 = 3,\ x_3 = 5,\ x_4 = 0$

14. Row-reduced columns 1, 5; $x_1 = 2,\ x_2 = x_3 = x_4 = 0,\ x_5 = 2,\ x_6 = 0$

16. Let a be the amount of A, b of B, c of C, d of D. In *standard form*:

$$\begin{aligned} &.2a && \le 200 \\ &.3a + .5b + .9c && \le 400 \\ &.1a + .2b + .1c + .1d && \le 50 \end{aligned}$$

$$a,\ b,\ c,\ d \ge 0$$

$$z = 4.45a + 3.35b + 2.35c + 4.80d$$

[The objective function is the simplification of:

$$\begin{aligned} z = 5a + 4b + 3c + 6d &- ((.2)(1) + (.3)(.5) + (.1)(2))a \\ &- ((.5)(5) + (.2)(2))b \\ &- ((.9)(.5) + (.1)(2))c \\ &- ((1)(1) + (.1)(2))d \ ] \end{aligned}$$

In *equation form*:

$$\begin{aligned} &.2a + s_1 && = 200 \\ &.3a + .5b + .9c + s_2 && = 400 \\ &.1a + .2b + .1c + 1d + s_3 && = 50 \end{aligned}$$

$$z = 4.45a + 3.35b + 2.35c + 4.80d$$

SECTION 9.2

2.

	c^*	x_1	x_2	x_3	x_4	x_5	x_6	x_7	x_8	**b**
x_2	1	2	1	1	0	3	0	-2	0	10
x_8	0	1	0	-1	0	1	0	1	1	9
x_6	0	1	0	1	0	2	1	2	0	7
x_4	1	3	0	2	1	0	0	-1	0	2
		3	0	6	0	3	0	-3	0	12

4.

	c^*	x_1	x_2	x_3	x_4	x_5	x_6	**b**
x_4	0	1	1	1	1	0	0	10
x_5	0	2	-1	-1	0	1	0	5
x_6	0	1	3	-4	0	0	1	1
		-2	1	-1	0	0	0	0

6.

	c^*	x_1	x_2	x_3	x_4	x_5	x_6	**b**
x_4	0	1	-3	-1	1	0	0	6
x_5	0	1	3	-4	0	1	0	8
x_6	0	1	-1	-3	0	0	1	1
		-1	1	-2	0	0	0	0

8. There is no minimal solution because the simplex indicator of column 1 is positive, but this column contains no positive entries.

10. There is no maximal solution because the simplex indicator of column 4 is negative, but this column contains no positive entries.

12. $x_1 = 0, \quad x_2 = 41/7, \quad x_3 = 29/7$

14. No maximum exists.

16. Produce 500 of toy D and no others.

SECTION 9.3

2.

$$\begin{aligned} x_1 + 2x_2 \qquad\qquad\qquad - s_1 \qquad\qquad + a_1 \qquad &= 6 \\ x_2 + x_3 + 4x_4 \qquad - s_2 \qquad\qquad + a_2 &= 12 \\ x_1 \qquad - 2x_3 + x_4 \qquad\qquad + s_3 \qquad\qquad &= 4 \end{aligned}$$

$$x_1,\ x_2,\ x_3,\ x_4,\ s_1,\ s_2,\ s_3,\ a_1,\ a_2 \geq 0$$

4.
$$\begin{array}{rrrrrrrrl} x_1 & + 2x_2 & - x_3 & & & & & & = 5 \\ & 4x_2 & - x_3 & + x_4 & + s_1 & & + a_1 & & = 8 \\ & -x_2 & + 3x_3 & - x_4 & & - s_2 & & + a_2 & = 2 \end{array}$$

$x_1, x_2, x_3, x_4, s_1, s_2, a_1, a_2 \geq 0$

Note: An artificial variable was not needed in the first constraint because of column 1.

6. $x_1 = 0, \quad x_2 = 12/7, \quad x_3 = 0, \quad x_4 = 18/7, \quad s_1 = s_2 = 0, \quad s_3 = 10/7$

8. $x_1 = 19/11, \quad x_2 = 26/11, \quad x_3 = 16/11, \quad x_4 = s_1 = s_2 = 0$

10. $x_1 = 0, \quad x_2 = 12, \quad x_3 = x_4 = 0;$ minimum value: 0

12. $x_1 = 0, \quad x_2 = 17/5, \quad x_3 = 9/5, \quad x_4 = 0;$ minimum value: 26/5

14. No maximum exists.

16. The amounts to be shipped from W_i to F_j are:

	F_1	F_2
W_1	0	0
W_2	2	4
W_3	5	0